AFRICAN INDUSTRIAL DESIGN PRACTICE

The underlying principle of this book is the African philosophy of Ubuntu, which acts as a guide for developing empathic products and services. The book makes the case that empathy is the key to any successful product and service design project because it enables designers to make wise design choices that align with users' demands.

Fifteen chapters provide the latest industrial design developments, techniques, and processes explicitly targeting emerging economies. At the outset, it covers the design context and the philosophy of the Ubuntu approach, which places people and communities at the centre of the development agenda. The book covers new product development, design research, design cognition, digital and traditional prototyping, bringing products to the market, establishing a company's brand name, intellectual property rights, traditional knowledge, and the business case for design in Afrika. It concludes with a discussion about the future of design and the skills aspiring designers will need.

African Industrial Design Practice: Perspectives on Ubuntu Philosophy will be an essential textbook for undergraduates, postgraduates, instructors, and beginner designers in emerging economies to provide regionally contextualised design processes, illustrated examples, and outcomes.

Richie Moalosi is Professor of Industrial Design at the University of Botswana. His research interests include design and culture, design research, design for sustainability, and social innovation. He is a member of the Institute of Engineering Designers, Pan Afrikan Design Institute, and Africa Design (Chapter of the Design Society).

Yaone Rapitsenyane is the Head of the Department of Industrial Design and Technology at the University of Botswana. He is also a service designer and a lecturer in sustainable design. His research interests include developing sustainable business models for SMEs and developing Sustainable Product-Service System curricula for African universities.

AFRICAN INDUSTRIAL DESIGN PRACTICE

Perspectives on Ubuntu Philosophy

Edited by Richie Moalosi and Yaone Rapitsenyane

Routledge
Taylor & Francis Group

LONDON AND NEW YORK

Designed cover image: © Midjourney – humanity towards others, interdependence, interconnectedness, and empathy

First published 2024Z
by Routledge
4 Park Square, Milton Park, Abingdon, Oxon OX14 4RN

and by Routledge
605 Third Avenue, New York, NY 10158

Routledge is an imprint of the Taylor & Francis Group, an informa business

British Library Cataloguing-in-Publication Data
A catalogue record for this book is available from the British Library

Library of Congress Cataloging-in-Publication Data
Names: Moalosi, Richie, editor. | Rapitsenyane, Yaone, editor.
Title: African industrial design practice : perspectives on Ubuntu philosophy/edited by Richie Moalosi and Yaone Rapitsenyane.
Description: New York : Routledge, 2024Z. | Includes bibliographical references and index.
Identifiers: LCCN 2023014433 (print) | LCCN 2023014434 (ebook) | ISBN 9781032218366 (hardback) | ISBN 9781032218311 (paperback) | ISBN 9781003270249 (ebook)
Subjects: LCSH: Industrial design – Africa. | Industrial design – Africa – Philosophy. | Ubuntu (Philosophy)
Classification: LCC TS115 .A37 2024Z (print) | LCC TS115 (ebook) | DDC 745.2096 – dc23
LC record available at https://lccn.loc.gov/2023014433
LC ebook record available at https://lccn.loc.gov/2023014434

ISBN: 978-1-032-21836-6 (hbk)
ISBN: 978-1-032-21831-1 (pbk)
ISBN: 978-1-003-27024-9 (ebk)

DOI: 10.4324/9781003270249

The Open Access version of chapter 2 was funded by University of Auckland.

CONTENTS

FIGURES

TABLES

CONTRIBUTORS

Samuel Oluwafemi Adelabu, Federal University of Technology, Nigeria. Samuel Oluwafemi Adelabu is a Senior Lecturer at the Department of Industrial Design, School of Environmental Technology at the Federal University of Technology, Akure, Nigeria. He has a background in Industrial Design (Ceramics) and holds a doctoral degree in Affective (Design) Science from the University of Tsukuba, Japan.

Angus Donald Campbell, University of Auckland, New Zealand. Angus Donald Campbell is the Director of Design and Deputy Head of Elam School of Fine Arts & Design, Faculty of Creative Arts & Industries, Waipapa Taumata Rau, the University of Auckland, Aotearoa, New Zealand. He is originally from South Africa. He holds a doctorate in development studies and a master's in industrial design.

Walter Chipambwa, Chinhoyi University of Technology, Zimbabwe. Walter Chipambwa is a Lecturer at Chinhoyi University of Technology in Zimbabwe. He has been a lecturer for the previous ten years. His interests include product design, computer-aided design, innovation, sustainable design, creativity, and design education and management. He is pursuing his PhD studies at the University of Botswana.

Patrick Dichabeng, Leeds University, UK. Patrick Dichabeng is a Lecturer of Product design at the University of Botswana. He holds a master's degree in Integrated Product Design. He specialises in computer-aided product design, rapid prototyping, design automation, human factors, and

acceptance. He is currently doing his PhD on human factors of highly automated vehicles.

Vikki Eriksson, Alto University, Finland. Vikki Eriksson is a researcher, con-
sultant, and design educator focusing on the relationship between design,
business, and technology. She holds a doctorate in Design. Her areas of interest include user experience, speculative practice, service design, and social
innovation practice in design.

Tawanda Gombiro, Botho University, Botswana. Tawanda Gombiro is a Lecturer at Botho University in the Faculty of Business and Accounting. He holds
a BCom (Hons) in Marketing and an MSc in Marketing and is pursuing a PhD
in Marketing. His research interests include sustainable marketing, customer
experience management, branding, service marketing, relationship marketing, and digital marketing.

Matlhogonolo Letsatsi, University of Botswana, Botswana. Matlhogonolo
Letsatsi holds an MSc in Manufacturing Management. He is a lecturer at the
University of Botswana. He has over 30 years of teaching, research and publi-
cation, development of curriculum, and community engagement experience.
He works with various community-based organisations to uplift and sustain
their livelihoods.

Odireleng Marope, University of Botswana, Botswana. Odireleng Marope is
a Communication Design Lecturer at the University of Botswana. He holds
a master's in Two-Dimensional and Digital Design. His research interest is in
communication design and cultural identity. He specialises in brand design
and development. He has over 15 years of teaching, research, and community engagement experience.

Richie Moalosi, University of Botswana, Botswana. Richie Moalosi is a
Professor of Industrial Design at the University of Botswana. His research
interests include design and culture, design research, design for sustainabil-
ity, and social innovation. He is a member of the Institute of Engineering
Designers, Pan Afrikan Design Institute, and Africa Design (Chapter of the
Design Society).

Shorn Molokwane, Botswana Institute of Technology, Research and Innovation, Botswana. Shorn Molokwane has been involved in creative practice, teaching, and research in design: aesthetics, user-centred design,
service design, fashion, and textiles for over 30 years. He heads the Divi-
sion of Design at the Botswana Institute for Technology Research and

Innovation. Shorn also runs institutional transformation through Design Thinking.

Olefile Bethuel Molwane, University of Botswana, Botswana. Olefile Bethuel Molwane is a Deputy Dean and Senior Lecturer at the University of Botswana. He has been the Head of the Department of Industrial Design and Technology at the University of Botswana. He has vast experience in teacher education and design and technology.

Thatayaone Mosepedi, University of Botswana, Botswana. Thatayaone Mosepedi is a Lecturer at the University of Botswana. He holds a master's degree in Product Design. He served as a teacher of Design and Technology for nine years at secondary schools. His research interests include product development, product management, future designs, design education, and additive manufacturing.

Keineetse Christopher Motlhanka, University of Botswana, Botswana. Keineetse Christopher Motlhanka is a Lecturer at the University of Botswana. He teaches courses in Computer-Aided Design, Design Studio, and Design Management. He holds an MSc in Product Design Engineering. He is pursuing his PhD in Design Processes for Additive Manufacturing at Loughborough University in the UK.

Tebo Motlhaping, Companies Intellectual Property Authority. Tebo Motlhaping is an Intellectual Property Expert based in Botswana. She works for the Companies and Intellectual Property Authority as a Copyright Specialist. She holds a master's in Intellectual Property. She has over eight years of work experience in technology and innovation, collective rights management, and general copyright administration.

Sekao Junior Motshubi, University of Botswana, Botswana. Sekao Junior Motshubi is a Lecturer at the University of Botswana. He holds an MSc in Educational Studies. He has a wealth of knowledge and experience spanning over 34 years, cutting across secondary and tertiary education institutions. His specialities include material technology and processes and model making.

Mugendi Kanampiu M'Rithaa, Machakos University, Kenya. Mugendi Kanampiu M'Rithaa is a transdisciplinary Industrial Designer/Educator/Researcher associated with several international design networks focusing on majority-world contexts and challenges. He is the Special Advisor to the Hasso Plattner School of Design Thinking at the University of Cape Town and President Emeritus of the World Design Organization.

Chinandu Mwendapole, Botho University, Botswana. Chinandu Mwenda-pole is a Lecturer at Botho University. She holds a PhD in Design and Intel-lectual Property laws. She has worked in academia for over 20 years and has experience in curriculum development, teaching, assessment, innovation for sustainable development, and quality assurance.

Sophia N. Njeru, Kirinyaga University, Kenya. Sophia N. Njeru, PhD (Design), is a fashion design lecturer and researcher at Kirinyaga University, Kenya. She has several years of university teaching and administration experience. She is avid in design for sustainability, particularly sustainable fashion in higher education institutions and industry. She is a member of the International Federation for Home Economics.

Matthews Ollyn, University of Botswana, Botswana. Matthews Ollyn is a Lecturer at the University of Botswana with 17 years of experience lecturing at the undergraduate level. Matthews is an industrial designer who specialises in user-centred design. His other area of interest includes strategic product design.

Caiphas Thusonyana Othomile, Oodi College of Applied Arts & Technology, Botswana. Caiphas Thusonyana Othomile is a Lecturer of Jewellery Design at Oodi College of Arts. He holds a Bachelor of Design (Design & Technology Education) and a diploma in Jewellery Design and Manufacture. He is also a practising jewellery designer who has won various national and international jewellery design competitions.

Debra Diana Ralitsha, Kwame Nkrumah University of Science and Technology, Ghana. Debra Diana Ralitsha holds a Doctor of Applied Art in Design from the Cape Peninsula University of Technology. She is a Lecturer in Communication Design at Kwame Nkrumah University of Science and Technology. Her research interests include social design, strategic design, service design, pluriversal design, social innovation, and sustainability.

Yaone Rapitsenyane, University of Botswana, Botswana. Yaone Rapitsen-yane is the Head of the Department of Industrial Design and Technology at the University of Botswana. He is also a service designer and a lecturer in sus-tainable design. His research interests include developing sustainable busi-ness models for SMEs and developing Sustainable Product-Service System curricula for African universities.

Victor Ruele, University of Botswana, Botswana. Victor Ruele is a Lecturer of Technology Education at the University of Botswana. He teaches mainly technology-related courses for pre-service and in-service teachers. Before this, he had worked as a teacher and head of the department in senior secondary schools and a teacher training college.

Oanthata Jester Sealetsa, University of Botswana, Botswana. Oanthata Jester Sealetsa is a Senior Lecturer in the Department of Industrial Design and Technology at the University of Botswana. His research interests include ergonomics and design, occupational health and safety, human-centred design, and technology transfer. He is a member of the Ergonomics Society of South Africa.

Polokano Sekonopo, University of Botswana, Botswana. Polokano Sekonopo holds an MSc in Mechanical Design and Theory from Northeastern University in China. He is a lecturer at the University of Botswana. His research interest includes assisting small micro-enterprises in building design capability. He is studying for a PhD in Design at the University of Botswana.

Keiphe N. Setlhatlhanyo, Leeds University, UK. Keiphe N. Setlhatlhanyo is a Lecturer and Researcher at the University of Botswana. She is currently a PhD researcher at the University of Leeds. Her interests include culturally significant designs and products, crafts, fashion/textile research, and community development. She has published numerous peer-reviewed journal articles and book chapters.

Patrick Sserunjogi, Kyamboko University, Uganda. Patrick Sserunjogi is a Designer, Educator, and Researcher at Kyamboko University. He holds a Doctor of Applied Art in Design form Cape Peninsula University of Technology. He is a lecturer of visual communication and design and multimedia. His research interest includes product design and development, user experience, creativity, and innovation.

FOREWORD

During my long stint with the World Design Organization (WDO), I have been fortunate to have had the opportunity to meet with brilliant people from around the globe – designers and design educators, government officials, young and old, male and female – who had an array of cultural and linguistic backgrounds. At one of the design events I was privileged to attend, my dear friend Prof. Mugendi Kanampiu M'Rithaa introduced me to a young and energetic professor, Richie Moalosi, from the University of Botswana, and we had the usual small talk. Subsequently, I learned that the University of Botswana was a member of our organisation, then called International Council of Societies of Industrial Design (ICSID), and has been supporting us actively for many years. Through my friendship with Prof. Mugendi, I became attracted to Afrika and visited South Africa in 2012. Later, when Prof. Mugendi got elected as the president of ICSID in 2013, I became very involved with Afrikan design events initiated by ICSID and later the World Design Organization (WDO). In later years, I got elected as the president of WDO, following in Prof. Mugendi's footsteps. I attended design events in a few other countries in Afrika, such as Kenya and Morocco. During these years, I also got to know more about Prof. Moalosi and his work on design in Botswana. When he asked me to write a foreword for his latest book, I was delighted to be part of his work on "Afrikan Design."

The editors begin their book entitled *Afrikan Industrial Design: Ubuntu Philosophy in Practice* by explaining the rationale behind using "k" instead of "c" in writing the word Afrika. This is an essential emotional factor for all Afrikans and is not well understood by people outside of Afrika. Even though Afrika's origins of design date back several thousand years, modern-day design has been credited to the Industrial Revolution and the years that followed. In their book, the editors discuss the value of human relationships in the design process. After several decades of design focus on materials and machinery, now the world is looking towards more human-centred design. In this context, this book is an appropriate contribution to that focus.

The editors also explain that the concept of 'Ubuntu' is often misquoted or misunderstood in the Western world. The editors emphasise that for Afrikan design to be culturally relevant and effective, design education should promote indigenous design methods tailored to local needs. This will allow designers to learn indigenous knowledge systems that will reflect in their work leading to benefiting a larger society than addressing a very narrow 'elite.'

The editors explain the complete product design process, from requirement gathering to getting the product out to consumers. The book is an ideal textbook for designers in Afrika who can benefit from the insights provided eloquently in several chapters.

I also enjoyed Chapter 2, in which the editors explain the concept of sustainability through a decolonisation lens and how design education in Afrika can influence designers with the idea and context of designing for sustainability in Afrika.

Another interesting topic in the book is ethics in design, which is well explained in Chapter 6. Drawing from the ethos of Ubuntu expressed through adages; the editors prescribe designers to self-regulate rather than be forced by others, which is very different from the Western philosophies.

The editors also provide details on how to handle the manufacturing of critical components which may not be able to follow the indigenous methods of manufacture, thus creating a fusion of practical processes with traditional processes.

The editors discuss designers' transformation from students to real business leaders in the final chapters. They also introduce the key constraints and opportunities for a bright Afrikan future for design and suggest that designers will become collaborators in a transdisciplinary world.

In summary, this book is an ideal guide for anyone who wishes to learn about the design practices specific to Afrika. The editors have done a marvelous job of bringing the value of culture and a human-centric approach to solving local design challenges.

Srini R Srinivasan
President Emeritus & Senator World Design Organization

ACKNOWLEDGEMENTS

The editors would like to express their most profound appreciation to the authors from various institutions of higher education in the following countries: Botswana, South Africa, Zimbabwe, Kenya, Uganda, Ghana, and Nigeria, who contributed chapters to this book. Each of the authors has provided a unique personal and professional contribution to the book under the spirit of sharing and generosity – tents of the philosophy of Ubuntu. This endeavour would not have been possible without your input. We also appreciate the reviewers who gave authors valuable feedback, which shaped this book. Your contributions will shape the practice of design in Afrika, which is based on the foundation of the philosophy of Ubuntu. There is hardly any industrial design book written by Afrikans to reflect the realities on the ground. This book contributes to the body of knowledge on the state of design in Afrika. The editors would not have managed to get this work to the readers without the continual support and vision of Matthew Shobbrook and Grace Harrison from Routledge. We want to express our deepest gratitude for your contribution.

PART 1

New Product Development

1

DESIGN CONTEXT IN AFRIKA

*Mugendi Kanampiu M'Rithaa, Richie Moalosi and
Yaone Rapitsenyane*

Introduction

Design in Afrika is still in its infancy compared to developed countries. There
are many challenges facing design education in Afrika, ranging from lack of
understanding from stakeholders to lack of recognition from government and
the public to lack of funding for design-related projects to weak enforcement
of intellectual property rights, etc. Despite all these challenges, design has
an immense scope in solving society's problems, especially those at the base
of the pyramid and challenges faced by small micro-enterprises and creative
industries, as these are the engines of economic growth in many developing
countries. Design should have an inclusive development approach that creates a society where citizens have equal access and opportunities for socio-
economic participation. Participation is not only about giving disadvantaged
individuals or groups a voice at the design table; it is about improving their
capacity to influence decision-making processes that will affect their lives.

Design can contribute to solving Afrikan challenges that include the
adverse impact of climate change, pandemics, increasing water scarcity,
desertification, biodiversity and ecosystem loss, low resilience to natural
disasters, energy crisis, food crisis, limited benefits from globalisation, the
global financial crisis, health security, trafficking and piracy, low penetration
of information communication technology services, urbanisation, developing
better disaster response mechanisms, genetically modified crops with regard
to food security and technology transfer, among others (United Nations Economic Commission for Africa, 2015). Afrika is still endowed with a wealth of
natural resources that the rest of the world cannot do without. However, raw
materials are exported to developed countries without any value addition.

DOI: 10.4324/9781003270249-3

The current design education needs to adequately contribute to developing the value chains of the exported raw materials.

Design education needs to rethink its approach using a decoloniality lens in addressing the outlined challenges. Design education in Afrika has been dramatically influenced by decolonisation. It needs to decolonise and promote the Afrikan voices in indigenising design education and developing culturally responsive design practices and processes. "Most successful creative people have all looked at the traditions, history and culture and then built on them" (The Impact of Culture on Creativity, 2009, p. 109). For example, design education in Afrika can infuse indigenous knowledge and the philosophy of Ubuntu into its practices, offering a new approach to tackling society's social problems for which current methods are deemed inadequate. This approach can assist in delivering new and sustainable lifestyles and building on cultural authenticity. In decolonising the design education in Afrika, this study has adopted the postcolonial theory to critique colonisation's oppressive educational and cultural practices. Postcolonial theory may help design educators and practitioners to reflect on their processes, practices, and contexts from a trans-cultural and global perspective.

To practise design in such an environment is a challenging task, and in most cases some of the methods and strategies used in the Global North need to be revised in Afrika. There is a need to contextualise such methods and strategies or develop new ones appropriate to the context. Afrikan designers need to look within their cultures to find out what has worked well within the society to solve problems that design can leverage. For example, the philosophy of Ubuntu has been successfully used to solve social challenges within societies for generations in Afrika. Design can leverage this known philosophy and develop new or improved design methods that leverage the ethos of the philosophy of Ubuntu which local people will easily relate to and understand. The chapter will also discuss the co-creation process from an Afrikan perspective, glocalisation, situating design in small micro-enterprises, design for the other 90% and principles for an effective design in Afrika.

The Decoloniality Lens

Global South cultures have been influenced by postcolonialism, and in the process, new cultural identities have emerged. Decolonising education and culture in the Global South involves challenging and rethinking the dominant Western values (Moalosi et al., 2017). Postcolonial theory is helpful in this study because it challenges specific patterns of domination by deconstructing the power relations and acknowledging the value of cultural membership. Design educators and practitioners in the Global South need to examine the effects of colonialism on curriculum transformation. Mbembe (2016) argues that Afrikan universities are local instantiations of a dominant academic model based on a Eurocentric epistemic canon. Thondhlana et al. (2021) advance that this has become hegemonic and actively represses anything articulated and envisioned outside this frame. Postcolonial theory is helpful to this book because it involves (a) the breaking down of Eurocentric codes, (b) recognising indigenous voices in the formation of postcolonial culture, and (c) recognising that the latter is, therefore, a hybrid culture.

Therefore, this is the time to change design education in Afrika, think differently about curriculum transformation, and start a new decolonised beginning (Becker, 2020). García and Baca (2019) critique postcolonial studies by claiming that coloniality still dominates the world today, hence the lack of transformation in the education sector. Ndlovu-Gatsheni (2019) argues that decoloniality is specifically formalised by the colonised in the Global South to dismantle global relations of power and conceptions of knowledge reproducing racial, gender, and geo-political hierarchies. The scholar advances that colonialism and coloniality still constitute the discursive landscape within which many forms of domination and exploitation are embedded.

Decoloniality is an alternative for resisting and analysing modernity, colonialism, and power's colonial matrix (Becker, 2020). Decoloniality is a vital tool for provoking long-established claims on higher education held by Euro-Americans, as they always dominated academic knowledge production and distribution, and weaker, poorer nations followed their education system without contextualising it (Majee & Ress, 2020). A deep understanding of the coloniality of knowledge enables Afrika to focus on teasing out epistemological issues, the politics of knowledge generation, and questions of who generates knowledge and for what purpose (Thondhlana et al., 2021; Mabvurira, 2020). Mignolo (2018) suggests that decoloniality can be achieved in several ways: protests, local and global movements, grassroots projects, and engagements, delinking from top-down hegemonic knowledge and linking to bottom-up pluriverse knowledge, powerful exploration of epistemological practices and curriculum to transform higher education. Decolonisation of the higher education curriculum could be addressed by adding new knowledge to the existing curricula and making visible the enunciation of knowledge and how it constructs curricula (Heleta, 2016). Begum (2015, p. 33) argues that postcolonial theory has practical application to design, which includes the following:

- It can produce further cross-cultural, anthropological knowledge that could have a crucial bearing on the design process and practice.
- It could help support cooperative and adaptive design methods towards user empowerment and design democratisation by breaking down hegemonic, hierarchical relationships between designers and end users.

Based on the abilities just listed, the authors subscribe to the epistemological lens that positions *Afrika* as an 'inside-out' perspective – the view of the continent espoused and co-created by its denizens, as a counterpoint of *Afrika*, which is an 'outside-in' perspective of how the rest of the globe views the continent and its myriad peoples.

Philosophy of Ubuntu

The concept of co-creation in design or other design methods can be explored from the ethical foundation of Ubuntu – a philosophy shared, under different names, by several Afrikan cultures. The philosophy is shared by numerous Bantu-speaking people throughout sub-Saharan Africa (Mabvurira, 2020). Ubuntu, translated from the Bantu language, means humanness, humanity towards others or *"I am because we are, and since we are, therefore, I am"* (Olasunkanmi, 2015). The Afrikan philosophy of Ubuntu defines a sense of self as one's relationships with other people shape it. It is based on humanness, caring, sharing, respect, compassion, and associated values that ensure a happy and qualitative community life in the spirit of family (Mabvurira, 2020). The perspective of human beings working collectively to achieve goals is infinitely more significant than any individual's potential. It is a way of living that starts with the assumption that 'I am' only because 'we are.' Ubuntu is rooted in an interpersonal form of personhood, meaning that one is because of the others. Ubuntu's emphasis on humanity and social relationships suggests that what is good for society is Ubuntu, and the contrary is true. This approach of equal humanity unlocks the door to complete tolerance and deep respect for the 'other' as an integrated gift that enriches one's society. The philosophy defines an activity for earning respect by first giving it and gaining empowerment by empowering others. In Ubuntu, relationships are valued instead of the individualistic form of identity. Afrikans believe that 'we can win together.' Ubuntu might serve as a counterbalance to the rampant individualism prevalent in Western society.

The philosophy embraces the idea that humans cannot exist in isolation. People depend on connection, community, and caring. It requires a conscious shift in how people think about themselves and others. A human being achieves humanity through their relations with other human beings. This consideration does not need to create an oppressive structure where the individual loses autonomy to maintain a relationship with an 'other.' Ubuntu is based on an absolute desire to see and harness other people's distinctiveness and variance, not as a threat but as a supplement to one's humanity, a common feature in design teams. The central pillar of the philosophy of Ubuntu, which has enabled Afrikan communities to survive centuries, is group solidarity. An individual in Afrika does not live in isolation but with others within a community. The philosophy is about people coming together and building a consensus around what social challenge affects the community. Once the community has debated on the challenge, what is best for the community is understood, and then one must buy into the consensus decision reached. Ubuntu promotes restorative justice and a community-centric ethos. For example, in Uganda there is '*Bataka kwegaita*' (communal solidarity), in Botswana there is '*boipelego*' (self-reliance), in Kenya there is '*Harambee*' (pulling together), in Tanzania there is '*ujamaa*' (familyhood), and in Ghana there is '*nobwa*' (mutual assistance). This philosophy has been used for centuries to mobilise sustainable development initiatives. There are valuable lessons the design profession can learn and adopt from such an indigenised sustainable development approach (M'Rithaa, 2009).

Ubuntu also extends to relationships with the non-human ecosystem (environment), such as rivers, plants, animals, etc. The assumption is that people cannot harm the fundamental things that society depends on, e.g., fauna and flora. Harming society is hurting oneself. Therefore, the philosophy of indigenous people, to which Ubuntu belongs, respects nature and sees nature as their extension. This communitarian ethos imposes a moral obligation regarding responsibility for others even before one thinks of oneself. Empathy and trust are built into the philosophy. Therefore, a measure of accountability is part of people's obligation, whether it comes to them through religion or a moral obligation of duty to others.

Consequently, Martin and Mirraboopa (2003, p. 11) state that "one experiences the self as part of others and that others are part of self; this is learnt through reciprocity, obligation, shared experiences, coexistence, cooperation, and social memory." These are values which are equally crucial for design for sustainability. The values espoused through the philosophy of Ubuntu include some. However, they are not limited to the following: generosity, respect, forgiveness, community spirit, honesty, cooperation, magnanimity, empathy, togetherness, ability to share, forgiveness, reconciliation, trust, empathy, collaboration, interdependence, spirituality, compassion, collective responsibilities, understanding, harmony, love, consensus building, and interpersonal relationships among people (Chilisa et al., 2016; Carroll, 2008). Eze (2010, pp. 190–191) sums up the philosophy of Ubuntu as:

> A person is a person through other people striking an affirmation of one's humanity through recognition of an 'other' in his or her uniqueness and difference. It demands a creative intersubjective formation in which the 'other' becomes a mirror (but only a mirror) for my subjectivity. This idealism suggests that humanity is not embedded in my person solely as an individual; my humanity is co-substantively bestowed upon the other and me. Humanity is a quality we owe to each other. We create each other and need to sustain this otherness creation. Moreover, if we belong to each other, we participate in our creations: we are because you are, and since you are, definitely I am. The 'I am' is not a rigid subject but a dynamic self-constitution dependent on this otherness creation of relation and distance.

Inclusive decision-making and participatory meetings are critical traditions in Afrikan communities' descendants. In seeking a solution for a problem, Afrikans share experiences, refer to the whole history of the society and consider the interests of the living and the dead. This procedure can be time-consuming, as it is carried on until consensus is achieved (Winschiers-Theophilus et al., 2012). To follow the philosophy of Ubuntu, designers need to identify themselves (as outsiders) as part of a broader society that includes designers from inside and outside who together derive a collective existence, and designers need to acknowledge that it is within this collective existence that "I am" a designer (Merkel et al., 2004). Winschiers-Theophilus et al. (2012) argue that to create new meanings about engagement, society outsiders must enter a lengthy process of social orientation. Designers should be aware of the intricacy of these encounters, reducing the process of reenvisioning the respective identities of the designers from the outside and inside contributes significantly to design failures in local communities.

The noble goals of the philosophy are significantly aligned with the co-creation, humanity-centred design, and participatory design processes. It is an inclusive and wide-ranging approach to design that considers the critical impact of design on society and the environment. The approach fosters products and services designed to be humanely produced, socially conscientious, environmentally regenerative, and supportive of communities and cultures. For example, in solving community challenges in Ubuntu, some elements of co-creation are used. The process empowers the community and recognises local people as a source of competence and experts in their experiences. Design can borrow some aspects from the Ubuntu philosophy to remain contextually relevant and co-create appropriate innovations. Embracing the spirit of Ubuntu into design thinking recognises non-Western ways of thinking rooted in Afrikan values and practices. It enhances more meaningful participation where socio-economic access to technological innovation and the epistemological conditions of designers and user groups differs acutely (Winschiers-Theophilus et al., 2012). Co-creation can lead to future pathways of value from which both the people and the enterprise can benefit. Though it is natural for people to want to use these for their self-progression and success, personal and societal benefits will be reaped if they use these qualities to better the community. The philosophy does not advocate sacrificing personal victory for the community's success, but people must balance the two; individual uniqueness is appreciated.

Nevertheless, one of the difficulties of practising the philosophy of Ubuntu in Afrika includes a need for more documentation compared to Western and Eastern ideologies. Afrikan indigenous knowledge is hardly documented, and in most cases, it is transmitted by storytelling from one generation to the other. However, efforts are underway to document the same. The young generation learns the philosophy of Ubuntu through direct interaction with elders in the local communities.

Co-Creation Process – an Afrikan Perspective

The authors have worked with several local communities and developed an indigenised co-creation process relevant to Botswana, Kenya, Uganda, South Africa, and other developing contexts. Cognisant of the need to decolonise the design process and infuse local indigenous knowledge, the philosophy of Ubuntu as a people-centred approach to solving challenges was incorporated into design practice. Ubuntu requires that individual values depend on cultural, social, and spiritual criteria. It requires an engagement with the community. Collectivism and communalism are crucial to the spirit of Ubuntu (M'Rithaa, 2009). The community is the foundation of all design activities, e.g., problem discovery, problem definition, and product/service design and development, and the process is participatory throughout. The design process starts within the community, and the outcome is a team effort of both the designer and the community. A spirit of solidarity concurrently strengthens cooperation and competitiveness amongst the team by enabling individuals to contribute their best efforts for their benefit. A team with increased members' loyalty, commitment, and satisfaction positively impact the overall design effort. The spirit of Ubuntu in design teams leads to collaborative and cooperative design efforts because the community is encouraged to participate, share, and support all the team members.

The relationship between the designer and the community is not based on the Western vertical power hierarchy structure but on a horizontal, communitarian system wherein all voices are accommodated and respected in the co-creation of knowledge. Subsequently, the community is not a passive recipient of knowledge constructed from the Western-influenced design process but a valued people who can co-create knowledge through the art of storytelling. The communities are experts in their experiences. The aim is to co-create innovative, effective, and sustainable solutions that are contextually relevant, culturally appropriate, and inclusive designs rooted in Afrikan values and the philosophy of Ubuntu. More details on the co-creation of the new product development process are explained in Chapter 3.

Global events impact design in Afrika; hence any design effort should selectively embrace such when designing products or services for the local context. This fusion in design has led to glocalisation.

Glocalisation

The evolution of human culture is characterised by exchanges, diffusion, hybridisation, and synthesis, where cross-breeding, borrowing, and adjusting to local needs are very common. Cultures in Afrika are not static, but other external factors from different cultures also influence them. Integrating the philosophy of Ubuntu in design brings to board the concept of *glocalisation*. Glocalisation is an amalgamation of the words *globalisation* and *localisation*.

The term describes a product or service developed and distributed for the global market but is also designed and customised to accommodate the local users' culture. That is, products and services should be suited to local tastes and interests yet have global application and appeal. The global market is open, and for Afrikan design to strive in the worldwide market, it will need to embed local users' cultural features, e.g., local features, customs, context, and needs or user preferences, and accommodate global interests and needs. It encourages designers from Afrika to 'design local and think global.' The 'design local and think global' initiative demands that designers understand users and their context.

Design teams must observe cultural practices and subtle nuances to understand the context deeply. This might include a society's history, religious beliefs, climate, geography, languages, aesthetics, and popular culture. Gorrie et al. (2018) argue that the socio-cultural perspective is explicitly included, absent from the conventional design, engineering, and business perspectives. The socio-cultural pathway emphasises development activities involving co-determination, knowledge transfer, and reliability of the product or service. Successful projects in a developing context are where the community or users have a say in the proposed solution. It has proved that co-creation or other participatory design methods lead to user engagement and ownership of the product, service, or system. Gorrie et al. (2018) outline the following socio-cultural deployment principles that need to be taken on board when designing products and services for a developing context:

- Observe how locals learn best and tailor the knowledge transfer process toward the same.
- Involve the stakeholders as much as possible so they feel a sense of ownership of the solution.
- Appreciate that community acceptance of the solution is crucial in any successful design.
- Maximise the unskilled hand fabrication/repair processes required by the stakeholders to limit risk exposure.
- Acknowledge the local perception towards repurposing materials to fix system failure points.

Adapting these principles could result in products and services designed around local contexts and motivations rather than being adapted or copied from other developed countries. Such a design experience may lead to products and services that match the needs and desires of the intended users in Afrika. Kang (2016) argues that design interventions have progressively taken on the form of grassroots activities. The design approach from Afrika seeks to move beyond merely empowering the local people but embraces an emancipatory agenda grounded in local participation and indigenous knowledge instead of relying on foreign products and services. Successful design practices are on situatedness and the agency of designers as facilitators, which results in the creation of common relationships between the people, the community, the socio-cultural context, technology, and artefacts, which together promote local empowerment/emancipation and sustainable development that meet the needs of the communities (Kang, 2016).

The social problems of the Global South are often complex and multilayered. Solving such challenges using a single linear method or one-off technical remedies is challenging. Transferring technology from the developed world is not necessarily a solution, and technologically well-developed projects often fail when deployed in developing contexts (Gorrie et al., 2018). The authors further argue that the fabric of infrastructure, materials, knowledge, and skills underpins the success of that technology in the developed country. This fabric does not necessarily exist to the same extent in the Afrikan case. Technology products and services can fail because they cannot be maintained due to a lack of materials, spare parts, or skills, or they are single-use and cannot be reconditioned. Most external non-governmental organisations, design companies, and other development organisations use the interventionist model to uplift Afrika. The upliftment of Afrika cannot be achieved simply by introducing technology from outside. There is a need to make sure that the products, services, and technologies are appropriate for the context at the current point of advancement and contribute sustainably to a course of upliftment.

There should be consideration of socio-cultural factors when designing with Afrikans. The introduction of new technology and products in society invariably changes the culture. Such changes are not always beneficial to the local communities. Designers have a moral obligation to implement technology and design products and services consistent with the community's culture, which necessitates a co-creation approach rather than the imposition of technology solutions (Gorrie et al., 2018). It is essential to work with the communities, learn their problem-solving methods and cultural customs, and use collaborative and participative design methods, which can assist in reaching sustainable, innovative, and appropriate solutions. For example, the authors have been using the co-creation process to brand products and services from community-based organisations ranging from indigenous coffee to beans to sweets to traditional ceramic products, etc. This approach will make technology interventions sustainable to the community's development agenda. Therefore, designers need to note that the local environment and socio-cultural dynamics should be considered when designing products and services rather than depending on the solution's financial and technical feasibility.

Kang (2016) cautions that challenges cannot be left to be solved by designers from more developed cultures in terms of socio-economic and political status. The outcome and spirit of outside designers will vanish soon after the design team has left. There will be weak continuity, hence the suggestion that there should be an elaborate exit strategy to avoid this scenario. Developing an exit strategy ensures sustainable outcomes that will stand the test of time. As proposed in the Ubuntu co-creation process, designers need to empower the participants from the beginning of the project, focusing on finding and reflecting on their indigenous knowledge, experiences, and values throughout the design process. The proven winning formula empowers local participation rather than relying on top-down training, which foreign non-governmental organisations, design companies, and development agencies often promote. Designers need to co-investigate real-world problems, immerse themselves in the context, actively participate in the phenomenon, step into the community, and co-experiment with a set of design practices (Kang, 2016). The anticipated vision is to have new products and services designed by Afrikans for Afrika and the world. This will create new jobs and manufacturing industries and open opportunities for the Afrikan youth.

Situating Design in Small Micro Enterprises

Small and micro enterprises In Afrika, we have many Small and Micro Enterprises than Medium Enterprises. Therefore, we have consistently used "Micro" throughout the book as they are the focus in our context. (SMEs) are the engine that drives world economies and industrial development and its consequent social benefits in the Global North and South (Muriithi, 2017; Iduarte & Zarza, 2010). SMEs play an essential role in most economies, particularly in Afrika. SMEs account for most businesses globally and are crucial contributors to job creation for the unskilled, semi-skilled, and skilled and for global economic development. They represent about 90% of companies and more than 50% of employment worldwide and contribute up to 40% of Afrika's national income (GDP) (World Bank, 2021). In Afrika, SMEs account for more than 90% of businesses and contribute about 50% to the GDP (Muriithi, 2017). However, these economically influential entities must benefit from good design practices (Carneiro et al., 2021). The scholars maintain that SMEs should use design for their strategic value in processes, operations, and strategy. This will make SMEs sustainable in the globalised context of social well-being (Carneiro et al., 2021).

The sector is significant due to its straightforward approach in response to most Afrikan needs by offering affordable products and services at reasonable terms and prices, besides being a source of income and employment and their capacity to innovate and create (Iduarte & Zarza, 2010). There is a niche for design in this sector since it is a creative and problem-solving profession. Good design is good business, as it can spark innovation, improve the stakeholder's experience, boost business growth, and increase profitability (Jewell, 2016). Afrika needs to invest in design to innovate its products and services and the know-how to protect its valuable intellectual property (IP) assets and leverage them in local and international markets. Despite its vast economic and innovative potential, the design potential is least recognised in Afrika.

For example, only some countries in Afrika have design policies and programmes. However, almost all countries have innovation policies driven by the Afrikan Union. Design is a driver of innovation. Therefore, there is a need to focus on design, and innovation will occur as part of the design process.

There are scanty systematic design and innovation programmes to instil a culture of continuous innovation and product development for SMEs in emerging markets (Austin-Breneman & Yang, 2017; Heikkilä & Heikkilä, 2017). This impedes the growth of the SME sector. SMEs in emerging economies are open to domestic and foreign competition, threatened mainly by cheap imports from other countries. The successful exploitation of new innovative ideas has driven the economic progress of many developed countries. Afrikan SMEs' challenge is competing based on unique value and innovation (Moalosi et al., 2013). Design and innovation matter because they mean higher quality and better products, more efficient services, and higher living standards. The absence of design and innovation can lead to business stagnation and the loss of jobs. Design has a significant role in assisting SMEs to innovate their products and services and remain globally competitive continuously.

The SMEs of today should be knowledge-based and a source of innovation because their success and survival depend on creativity, innovativeness, and design. SME businesses are facing new challenges in increasing their creativity and innovation rate. Innovation is vital for SMEs to survive and grow in the long term. Design-driven SMEs are more innovative than those which do not embrace design and can survive global competition (Lawlor et al., 2015). They often greatly tolerate higher-risk initiatives and value new ideas and originality. The challenge for SMEs is to move from competing with a relatively low cost of doing business to using design and innovation as competitive business tools. Afrikan SMEs can only compete in this new environment if they become more innovative and embrace design in their practice. Design should be viewed as a strategic business tool that can be used to develop innovative products and services. A business that is not growing through introducing new products and services is likely to decline. However, SMEs' effective use of design can positively contribute to business performance and competitiveness (Moalosi et al., 2013; Lawlor et al., 2015; Heikkilä & Heikkilä, 2017).

There is a need to promote awareness and design-led strategies supported using IP rights among SMEs. To demonstrate the value of design to SME businesses, the World Intellectual Property Organisation rolled out a project in Morocco to create awareness and understanding of the advantages of design supported by the effective use of IP rights. It was discovered that there were low levels of IP awareness, and the economic potential was largely untapped and ripe for development (Jewell, 2016). The SMEs involved in the project produced innovative designs legally protected at the end of the exercise. This shows that design can contribute to a remarkable transformation in SMEs' business success. Well-designed glocalised products and services stand out in the market and translate into greater demand, increased profits, and job opportunities. There is a scope to use design to solve Afrika's pressing challenges, such as water supply, food security, access to clean energy, youth unemployment, climate change, sanitation, and gender parity.

Designers in Afrika should focus on more than just market pull as a driver of design innovation. They should also consider technology push as another driver of design innovation. The market-pull approach involves stakeholders co-creating incremental innovation. However, this may be detrimental to radical innovation (Scaringella et al., 2017). Heikkilä and Heikkilä (2017) argue that when an enterprise focuses on existing stakeholders, it may not recognise opportunities in emerging markets that disruptive solutions may serve. Therefore, the process driven by technology push often leads to radical innovations, whereas market pull is more often restricted to incremental innovations (Heikkilä & Heikkilä, 2017). SMEs with little design experience can outsource design services outside the enterprise in product innovation. External design experts can introduce new ideas, significant innovation and creativity, and solutions the enterprise could not afford (Carneiro et al., 2021). This approach will help introduce design in SMEs and help enterprises recognise the strategic value of design in product and service development. Kootstra (2009) and Carneiro et al. (2021) propose that SMEs should integrate design at four levels, thus:

- *Level 1* (no design): Internal expertise on design methods, tools, and processes needs to be present, and this produces uncertain results.
- *Level 2* (project): Design is restricted to aesthetics/styling, product line extensions, and product improvement. At this stage, design is mainly overlooked as a tool for creating added value.
- *Level 3* (function): At this stage, the priority is optimised design processes and high-quality outcomes across the enterprise. The enterprise has a devoted design function.
- *Level 4* (culture): Design actively develops, supports, and optimises enterprise strategy.

Due to the limited design knowledge in Afrika, most enterprises are at level 1 (no design). The literature shows that effective use of design can positively contribute to SMEs' business performance (Carneiro et al., 2021). SMEs must learn to use design as a strategic resource for their entities.

Principles for an Effective Design in Afrika

In addition to the socio-cultural deployment principles outlined by Gorrie et al. (2018), Mattson and Wood (2014) advance nine principles that can support the design of products and services in Afrika as follows:

1 *Co-creating with people from the specific developing context encourages designer empathy, promotes stakeholder ownership, and empowers*

individuals. Designers must immerse themselves and interact with local stakeholders in the socio-cultural context for which the product or service is intended. The stage should be approached with an open mind, as there will be valuable lessons learnt in the process that will inform the design process.

2 *Multiple testing of the product or service in the actual socio-cultural context is vital for Afrika design*. Testing the product or service in the problem context will assist the design teams in understanding if stakeholders are using the product or service the way it is intended to be used and assess whether the product or service meets their needs. This stage should not be a post-development activity but an ongoing collaborative process throughout the product development stages. This will enhance trust issues and ownership of the product or service.

3 *Imported technologies should be adapted to the specific developing context*. If technologies are contextualised, they will be effective and sustainable. It is vital to understand the social context of the stakeholders and only implement technologies that integrate well into Afrika.

4 *Design products and services to alleviate poverty to benefit individuals and communities in urban and rural contexts*. The socio-cultural contexts in rural and urban areas are unique and different. This influences the kind of product or service to be designed. The designer can consider designing with the urban, rural, or both stakeholders to have a sustainable design with the maximum impact.

5 *Poverty alleviation efforts affect more women and children than men*. Deliberate efforts should include women and children in the co-creation design process at the community level. This will ensure that products and services are designed to their exclusive needs.

6 *Adapt project management techniques specific to the Afrikan context to enable a more effective and efficient design process*. Intentionally select, adapt, and implement a customised project management strategy that will articulate responsibilities and accountability and use different communication styles. Designers should not impose any risk and project management tools on stakeholders.

7 *Products and services designed for Afrika significantly impact when interdisciplinary teams contextualise, develop, and implement*. Designers should build multi-disciplinary and interdisciplinary teams, including the community, to address social, political, ecological, and economic challenges. Such an approach will be based on Ubuntu's ethos, which seeks collective ownership of ideas and responsibility.

8 *Collaboration with the government, local authorities, and influencers contextualises the design and facilitates poverty alleviation plans*. A scoping exercise should be conducted before any design activity can take place. The exercise enables designers and the community to know each other, and it is a building block of trust and project buy-in. One must go through gatekeepers in rural areas to introduce them to local authorities, e.g., the

village chief, councillor, or local influencers. If this vital step is missed, the project will fail because of a lack of buy-in and proper consultation. If the local authorities and their communities have given the project the green light, the chances of making a successful project are very high.

9 *Use existing distribution strategies to successfully introduce products and services into the Afrikan markets.* Designers can research the successful distribution strategies of similar products in their specific settings and contextualise the same.

Conclusion

This chapter argues for an Afrikan perspective on the increasingly popular design methodology. At its core, design thinking is a practical application of nuanced and context-responsive humanity-centred design strategies deployed in a participatory design context to aid problem-solving and result in innovative solutions. Within the Afrika context, design is a relatively new approach gaining traction as more design practitioners and participating actors realise the efficacy of such an ethos. The value of design is well received within homogenised corporate team settings, applying the methods and toolkits in financially constrained environments (such as those in Afrika and other majority world contexts). This poses unique challenges that demand interrogation, empathy, and a deeper understanding to unlock latent creative and innovative potentialities within participating communities.

Further, the chapter calls for reflection and introspection on the approach design takes in Afrika. Design education has been heavily influenced by colonialism and globalisation. It is an opportune time to decolonise design education and encode local thoughts such as indigenous knowledge and the philosophy of Ubuntu. Postcolonial theory is needed to challenge some assumptions accepted as accurate and how design knowledge is generated. The chapter discussed how the philosophy of Ubuntu could be linked to design education and profession. The philosophy of Ubuntu advocates for a humanist, spiritual, consensus-building, and dialogue- and community-based society. The community is more significant than the individual. The philosophy of Ubuntu represents a prevalent spirit of caring and humanness within the community. Anybody who does not recognise themselves in the community is considered an outsider, which conflicts with Western ideologies. Therefore, any design efforts should be aligned with the ethos of Ubuntu.

The socio-cultural deployment principles and the nine principles interrogated herein have proven efficacy in supporting the design of products, services, systems, and stakeholder experiences in Afrika – they indicate a departure from the developed world's traditional approach to design practice. Consequently, developed countries' design tools and strategies may not work in emerging economies. To ensure usefulness, relevance, and efficacy in the developing context, there is a need to contextualise the design methods co-created in industrially developed countries. It is worth borrowing from the anthropocentric ethos of the philosophy of Ubuntu as it has been widely used in Afrika to solve socio-economic challenges. Socio-cultural factors and context play a leading role in developing products and services in Afrika. Design in Afrika needs to approach challenges from a glocalised perspective. This will enable SMEs to use design as a competitive tool for innovation. It will also allow SMEs to unlock the continent's vast creative and entrepreneurial potential, leading to the design of products, services, systems, and user experiences that are locally relevant and globally appealing. A deep understanding of users and their context may lead to innovative, contextualised, and sustainable design solutions contributing to the United Nations Sustainable Development Goals. The Ubuntu approach, coupled with emerging virtual co-design platforms and online tools (such as Sprintbase™, Mockplus™, Mural™, Miro™, and InVision™), could play a pivotal role in shaping the success of any empathic design efforts diverse to the Afrikan context that recognise the communities' socio-cultural values.

Reference List

Austin-Breneman, J., & Yang, M. (2017). Design for micro-enterprise: A field study of user preference behaviour. *21st International Conference on Engineering Design*, ICED17, pp. 21–25 August 2017, The University of British Columbia, Vancouver, Canada.

Becker, A. (2020). Decolonial human rights education: Changing the terms and content of conversations on human rights. *Human Rights Education Review*, 5. http://doi.org/10.7577/hrer.3989.

Begum, T. (2015). *A Postcolonial Critique of Industrial Design: A Critical Evaluation of the Relationship of Culture and Hegemony to Design Practice and Education Since the Late 20th Century*. PhD thesis, University of Plymouth. https://pearl.plymouth.ac.uk/handle/10026.1/3410 [Accessed 19 October 2021].

Brown, T. (2019). *Change by Design, Revised and Updated: How Design Thinking Transforms Organisations and Inspires Innovation*. Boston: HarperCollins.

Carneiro, V., da Rocha, A. B., Rangel, B., & Alves, J. L. (2021). Design management and the SME product development process: A bibliometric analysis and review. *She Ji: The Journal of Design, Economics, and Innovation*, 7(2), pp. 197–222. https://doi.org/10.1016/j.sheji.2021.03.001.

Carroll, K. K. (2008). Africana studies and research methodology: Revisting the centrality of African worldview. *Journal of Pan African Studies*, 2(2), pp. 5–27.

Chilisa, B., Major, T. E., Gaotlhobogwe, M., & Mokgolodi, H. (2016). Decolonising and indigining evaluation practice in Africa: Towards African relational evaluation approaches. *Canadian Journal of Program Evaluation*, 30(3), pp. 347–362.

Eze, M. O. (2010). *Intellectual History in Contemporary South Africa*. Basingstoke: Palgrave.

García, R., & Baca, D. (2019). Introduction. Hopes and visions: The possibility of decolonial options. In R. García & D. Baca (Eds.), *Rhetoric Elsewhere and Otherwise. Contested Modernities. Decolonial Visions*, pp. 1–46. Champaign IL: NCTE.

Gorrie, A. W. T., Pons, D. J., Maples, D., & Docherty, P. D. (2018). Principles of product design in developing countries. *Applied System Innovation*, 1(2), p. 11. https://doi.org/10.3390/asi1020011.

Heikkilä, J., & Heikkilä, M. (2017). Innovation in micro, small and medium-sized enterprises: New product development, business model innovation and effectuation. *30th Bled eConference: Digital Transformation – From Connecting Things to Transforming Our Lives*, Bled, Slovenia. http://doi.org/10.18690/978-961-286-043-1.15.

Heleta, S. (2016). Decolonisation of higher education: Dismantling epistemic violence and Eurocentrism in South Africa. *Transformation in Higher Education*, 1(1), a9. https://doi.org/10.4102/the.v1i1.9.

Iduarte, J. T., & Zarza, M. P. (2010). Design management in small- and medium-sized Mexican enterprises. *Design Issues*, 26(4), pp. 20–31.

The Impact of Culture on Creativity: A Study Prepared for the European Commission (Directorate-General for Education and Culture) (2009). www.keanet.eu/docs/impactculturecreativityfull.pdf [Accessed 16 October 2021].

Jewell, C. (2016). Unlocking design potential in developing countries. *WIPO Magazine*. www.wipo.int/wipo_magazine/en/2016/03/article_0002.html.

Kang, L. (2016). Social design as a creative device in developing countries: The case of a handcraft pottery community in Cambodia. *International Journal of Design*, 10(3), pp. 65–74.

Kootstra, G. L. (2009). *The Incorporation of Design Management in Today's Business Practices. An Analysis of Design Management Practices in Europe*, trans. Erwin Postma. Rotterdam: The Hague and INHOLLAND University. http://Lastrategiedesign.com/public/DME_Survey09.pdf.

Lawlor, P., O'Donoghue, A., Wafer, B., & Commins, E. (2015). *Design-Driven Innovation: Why it Matters for SME Competitiveness*. Dublin: Plus Print Ltd.

Mabvurira, V. (2020). Hunhu/Ubuntu philosophy as a guide for ethical decision making in social work. *African Journal of Social Work*, 10(1), pp. 73–77.

Majee, U., & Ress, S. (2020). Colonial legacies in internationalisation of higher education. *A Journal of Comparative and International Education*, 50, pp. 463–481.

Martin, K., & Mirraboopa, B. (2003). Ways of knowing, being and doing: A theoretical framework and methods for indigenous and indigenist re-search. *Journal of Australian Studies*, 27(76), pp. 203–214. http://doi.org/10.1080/14443050309387838.

Mattson, C. A., & Wood, A. E. (2014). Nine principles for design for the developing world as derived from the engineering literature. *Journal of Mechanical Design*, 136(12), MD-13-1442. https://doi.org/10.1115/1.4027984.

Mbembe, A. (2016). Decolonising the university. *Arts and Humanities in Higher Education*, 15(1), pp. 29–45.

Merkel, C., Xiao, L., Faroog, U., Ganoe, C., Lee, R., Carroll, J., & Rosson, M. (2004). Participatory design in community computing contexts: Tales from the field. *Paper Presented at the 8th Biennial Participatory Design Conference*, Toronto, Canada.

Mignolo, W. D. (2018). What does it mean to decolonise? In W. D. Mignolo & C. E. Walsh (Eds.), *On Decoloniality*, pp. 105–134. Durham: Duke University Press.

Moalosi, R., Marope, O., & Setlhatlhanyo, K. N. (2017). Decolonising Botswana's design education curriculum by infusing indigenous knowledge: Botho co-creation process. In M. T. Gumbo & V. Msila (Eds.), *African Voices on Indigenisation of the Curriculum: Insights from Practice*, pp. 66–96. Wandsbeck, South Africa: Reach Publishers.

Moalosi, R., Molokwane, S., Sealetsa, O. J., Molwane, O. B., Letsholo, P., Letsatsi, M., & Mwendapole, C. (2013). *Small Micro Enterprises Landscape in Gaborone and Surrounding Areas and Their Readiness for Product Innovation*. Gaborone International Design Conference. www.researchgate.net/publication/264547399_Small_Micro_Enterprises_landscape_in_Gaborone_and_surrounding_areas_and_their_readiness_for_product_innovation.

M'Rithaa, M. K. (2009). Embracing sustainability: Revisiting the authenticity of 'event' time . . . *Proceedings of the 2nd International Symposium on Sustainable Design (ISSD)*, 5–6 November 2009, Sao Paolo.

Muriithi, S. M. (2017). African Small and Medium Enterprises (SMEs) contributions, challenges and solutions. *European Journal of Research and Reflection in Management Sciences*, 5(10), pp. 36–48.

Ndlovu-Gatsheni, S. J. (2019). Discourses of decolonisation/decoloniality. *Papers on Language and Literature*, 55(3), pp. 201–226.

Olasunkanmi, A. (2015). Euthanasia and the experiences of the Yoruba people of Nigeria. *Ethics & Medicine*, 31(1), pp. 31–38.

Scaringella, L., Miles, R. E., & Truong, Y. (2017). *Customers' Involvement and Firm Absorptive Capacity in Radical Innovation: The Case of Technological Spin-offs*. Technological Forecasting and Social Change. www.researchgate.net/publication/312355628_Customers_involvement_and_firm_absorptive_capacity_in_radical_innovation_The_case_of_technological_spin-offs.

Thondhlana, J., Abdulrahman, H., Garwe, E. C., & McGrath, S. (2021). Exploring the internationalisation of Zimbabwe's higher education institutions through a decolonial lens: Postcolonial continuities and disruptions. *Journal of Studies in International Education*, 25(3) 228–246. https://doi.org/10.1177/1028315320932319.

United Nations Economic Commission for Africa (2015). *New and Emerging Challenges in Africa Summary Report*. https://sustainabledevelopment.un.org/index.php?page=view&type=400&nr=502&menu=1515#:~:text=s%20challenges%20include%20the%20adverse,limited%20benefits%20from%20globalization%2C%20health.

Winschiers-Theophilus, H., Bidwell, N. J., & Blake, E. (2012). Community consensus: Design beyond participation. *Design Issues*, 28(3), pp. 89–100.

World Bank (2021). *Small and Medium Enterprises (SMEs) Finance*. www.worldbank.org/en/topic/smefinance.

2

ADVANCING AFRIKAN SUSTAINABLE DESIGN

Angus Donald Campbell and Yaone Rapitsenyane

Introduction

The United Nations Human Development Report (2020) highlighted that "broken societies [have] put people and planet on [a] collision course." A year later, the Intergovernmental Panel on Climate Change (IPCC) report on the Physical Science Basis (Masson-Delmotte et al., 2021) made humanity's 'code red' even more apparent. Based on the outcomes of the last report, the Secretary-General of the United Nations (UN), António Guterres, highlighted that "The alarm bells are deafening and the evidence is irrefutable: greenhouse gas emissions from fossil-fuel burning and deforestation are choking our planet and putting billions of people at immediate risk" (IPCC, 2021).

A significant reason for humanity's current mess is the constant growth agenda of Western capitalism, which ignores inherent planetary resource limits (Meadows et al., 2004). Such a world view selfishly places humans at the top of the food chain instead of understanding its complex interconnection. The scale of the human impact on the earth has led many scientists to accept the description of the contemporary era as the 'anthropocene' (Crutzen & Stoermer, 2000) – an era in which the impacts of human activity are evident at almost every level of the planetary ecosystem.

In this chapter, the authors will contextualise how the Global North has responded to unsustainable development, focusing on the Sustainable Development Goals (SDGs). The parallel shift in industrial design will be explored from the perspective of the Global North. This will be contrasted with the Afrikan concept of Ubuntu to explore a more endogenous conception of sustainable design in Afrika. Finally, case studies will be discussed with examples of how Afrikan sustainable design is attempted in contemporary Afrikan industrial design.

DOI: 10.4324/9781003270249-4

Sustainable Development and Design

The definition of human development has been described as being at loggerheads with the natural world. In his seminal book *The History of Development*, Gilbert Rist (1999, p. 13) describes it as:

> a set of practices, sometimes appearing to conflict with one another, which require – for the reproduction of society – the general transformation and destruction of the natural environment and social relations. It aims to increase the production of commodities (goods and services) geared, by way of exchange, to effective demand.

Rist (1999, p. 19) notes that such a definition is "scandalous" because it seemingly contradicts the belief that development is focused on enhancing humanity's equality. However, when the origins of development are contextualised as a post-war drive to further the markets of the Global North (Truman, 1949), it becomes clear that such a project did not derive from altruistic intentions. With a narrow focus on economic growth, the initial development measurements were crudely measured by gross domestic product (GDP). It was only after the report from the Brundtland Commission in 1987 that sustainability and broader conceptions of human well-being were identified as additional metrics for human development.

The Brundtland Report (United Nations, 1987, p. 41) defined sustainable development as "development that meets the needs of the present without compromising the ability of future generations to meet their own needs." The Brundtland definition further unpacks these needs in two concepts: firstly, the priority of meeting the basic needs of the world's poorest and, secondly, the immense pressure technological advancement and social organisation have placed on the environment's ability to sustain present and future needs. These concepts of sustainable development sought to redress human activity misaligned with the natural ecosystem to regenerate from the uncontrolled extraction of resources.

The United Nations Department of Economic and Social Affairs (2022) alludes that the 17 Sustainable Development Goals (SDGs) are considered one of the most critical metrics for the world's progress towards environmental sustainability. However, with what seems to be a hang-on from the original conception of development, many of the nations with the highest SDG scores in the index – the most 'developed' in terms of the socio-economic and political characteristics in the Global North – also have excessive and highly unsustainable material consumption practices (Hickel, 2020). Bhutan and Suriname are the only carbon-negative countries (Goering, 2021). This is primarily due to their small populations and a large proportion of forests. However, despite their stellar environmental standing, they are considered 'underdeveloped' countries in the Global South. All nations are developing despite the binary of developed and underdeveloped. However, the indices that measure sustainable development bring biases of the past.

The same bias is experienced in the development of design (Campbell, 2013). Industrial design as a discipline also originated in post-war Global North economies, where its focus was also economical, specifically focusing on increasing production and consumerism of products through enhancing function and aesthetics. In the opening of his book *Design for the Real World: Human Ecology and Social Change*, seminal designer and educator Victor Papanek notes, "There are professions more harmful than industrial design, but only a few" (1971/1985).

Since the 1980s, the discipline of industrial design has undertaken two significant shifts. The first was a move to a more human-centred design approach that was less focused on creating consumers but on meeting real human needs by designers who acknowledged that they were not necessarily the experts on others' lived experiences. There was also an acknowledgement that industrial design had moved from being predominantly solution-oriented to becoming strategic as it includes a more comprehensive range of outcomes ranging from products to services, systems and experiences (World Design Organization [WDO], 2015).

However, as per the human-centred descriptor, this approach to industrial design still hierarchically tended to serve the interest of privileged people over the planet. This led to the next important and more recent shift towards earth-centred approaches to design. This shift acknowledges design interventions' complex and interconnected context from a socio-technical systems perspective, including natural ecologies. The seminal book *Design for Sustainability: A Multi-level Framework from Products to Socio-technical Systems* by Ceschin and Gaziulusoy (2020) explores the evolution of sustainable design from green design to design for sustainability transitions. Ceschin and Gaziulusoy (2020) present a comprehensive overview of the development of sustainable design from its insular and technocentric beginnings to a more systemic and earth-centric future. As an open-source published book, the book provides a detailed resource that attempts to document the development and range of methods used in sustainable design in the Global North. Some of the critical approaches to design for sustainability are discussed next.

Sustainable Product Design

Most conceptions of sustainable product design are focused on the entire life cycle of a product, from its natural resource until its final demise, or in more considered conceptions, into a new life as something else (McDonough & Braungart, 2002). These approaches are described under green design, eco-design, whole life cycle, or cradle-to-cradle. Some are more authentic than others, with the idea of green-washing pervading many supposedly sustainable products using what seem to be natural materials or green or eco-branding. For a more authentic approach to sustainable product design, designers must carefully consider and measure all the resources, manufacturing processes, potential deconstruction and remanufacture in a product's entire life-cycle (Cradle to Cradle Products Innovation Institute [CCPII], 2022). As an example of cradle-to-cradle thinking, McDonough and Braungart (2002) return to nature in their description of a tree-growing fruit cycle. The fruit will provide sustenance for other animals before being returned to the soil, decomposing to its constituent molecules without degrading them. It will make them fully available as compost for the tree seeds or for other life to grow. Energy is used in the fruits' creation through light and nutrients from the soil. Still, nothing becomes unusable waste or harms the environment in this life cycle of making and decomposition.

The main problem with a product-centric focus for design is that products tend to be resource- and energy-intensive. Despite the best attempts at design for disassembly, reuse or recycling, many of the reimaginings of a product tend to result in a less valuable or useable form than their original form. Another problem with a product-centric focus is that many products are already available worldwide. The problems faced today, particularly the intractable ones, do not necessarily need more products to 'solve' them. This is where a service-oriented approach to design can add a more sustainable alternative.

Service Design for Sustainability

Service design is increasingly becoming important for organisations wishing to transform or improve their customers' services. Service design is defined as the "design for experiences that happen over time and across different touch-points" (Service Design, n.d.). Touchpoints are crucial in designing services, as they are the points of interaction between customers and service providers (Stare & Križaj, 2018). A shift from product to service and experience inherently limits a need for more resources. It also allows existing products to be more pleasurable or used more efficiently.

A focus on service experiences by society puts pressure on manufacturing companies that focus on product development to re-direct their innovation activities towards service-oriented differentiation. This is analogous to the shift from a manufacturing to a service economy (Vargo & Lusch, 2004). A service differentiation strategy for manufacturing companies means the value is defined less in tangible terms and more in intangible and dynamic services produced and consumed simultaneously (Tukker, 2015; Vargo & Lusch, 2004). This strategy makes services the core offering supported by enabling products rather than being add-ons to products, as in traditional product-oriented strategies (Gebauer et al., 2016). Service differentiation depends on the capabilities companies develop over time with their stakeholders (Bello et al., 2016). A stakeholder relationship approach also means that revenue can be generated at different life cycle stages of the offering (Manzini & Vezzoli, 2003; Tan & McAloone, 2006).

Design for Social Innovation

Social value can be created by solving social problems through new ideas that work at meeting social goals (Porter & Kramer, 2019). All human needs have a social dimension to them. This includes needs met through profit-making ventures. Phills et al. (2008) defined social innovation as

> any novel and useful solution to a social need or problem that is better than existing approaches (i.e., more effective, efficient, sustainable, or just) and for which the value created (benefits) accrues primarily to society as a whole rather than private individuals.

This definition implies that creativity is required for solutions to improve on existing ones. The edge in social innovations emanates from the co-creation of value by all stakeholders for mutual social benefit rather than profit – the driver to innovate is social needs rather than opportunities to make money.

Socio-technical Systems Design

Product-service systems (PSS), also widely discussed alongside servitisation (Baha et al., 2014; Baines & Lightfoot, 2013; Morelli, 2003; Vezzoli et al., 2021), can be viewed as an integration of new product development and new service development (De Lille et al., 2012). By simultaneously addressing product and service components of value creation, PSS aims to shift the business focus from designing (and selling) physical products to designing (and selling) a system of products and services which are jointly capable of fulfilling specific client demands while re-orienting current unsustainable trends in production and consumption practices (Manzini & Vezzoli, 2003).

PSS is a business strategy based on continuous life cycle improvement, considering the product and service life cycles (Kjaer et al., 2019; Tan & McAloone, 2006). In this way, the concept represents a holistic approach to sustainable innovation. Through this strategy, manufacturing companies can undergo servitisation to redefine value creation in non-product terms (Rapitsenyane et al., 2019). According to Tomiyama (2001), the value of this process of servitisation is in intensifying service contents of offers to arrive at the environmentally conscious design and manufacturing and create more added value in future advanced societies. A view of the whole landscape of the problem, the environment in which the problem is being investigated, relationships between factors causing the problem and possible factors that might lead to a solution is necessary for this holistic view, especially if looked at from the design perspective. A whole system design approach is necessary to aid such decisions (Fiksel, 2006) and move design away from its traditional focus on material products (Morelli, 2003). The position of PSS in a systemic context can be articulated in terms of the tangible and intangible value that requires an environment, provider, consumer and product to facilitate its provision (Tomiyama, 2001).

An Afrikan Concept of Sustainability

These brief prior explorations of sustainable design arise from a hierarchical global position from the Global North. Development and the indices that measure it do not account for the fact that most 'developed' nations could achieve their position in the world on the back of the labour and resources of their colonies. Latin American and decolonisation scholar Nelson Maldonado-Torres defines decoloniality as "the dismantling of relationships of power and conceptions of knowledge that foment the reproduction of racial, gender, and geo-political hierarchies that came into being or found new and more powerful forms of expression in the modern/colonial world" (2006, p. 117).

Colonisation and the political, social and economic systems that arose from it consciously undermined indigenous Afrikan cultures. Many Afrikan societies have pre-colonial oral traditions with a deep cultural appreciation for the interdependence of a person's physical well-being, the well-being of the environment, the community (past and present) and spiritual factors beyond the physical realm (Kideghesho, 2008). The Nguni Bantu concept of Ubuntu acknowledges these more comprehensive relations, connections, and responsibilities as,

> A collection of values and practices that people of Africa or African origin view as making people authentic human beings. While the nuances of these values and practices vary across different ethnic groups, they all point to one thing – an authentic individual human being is part of a larger and more significant relational, communal, societal, environmental, and spiritual world.
>
> *(Mugumbate & Chereni, 2020, p. vi)*

The separation between people and the planet is evident in the Global North (Eisenstein, 2013). Indigenous world views such as Ubuntu present a far more sustainable, integrated and communal conception of human relations and relationships with the natural environment (Katz, 1937/2011; Ogude, 2019; Zondi, 2021, pp. 237–238). Ubuntu, as a philosophy, is ultimately focused on the essence of humanity through our relations and connectedness to others. *Umuntu ngumuntu ngabantu* translates from isiZulu as 'a person is a person through other people.' The recently deceased Desmond Tutu (2012, pp. 34–35) once described someone who displayed Ubuntu as one who understood that "my humanity is caught up, is inextricably bound up, in theirs." Although Ubuntu as a concept is never directly mentioned, Bowles and Gintis's (2011) book *A Cooperative Species: Human Reciprocity and Its Evolution* explores the significant effect that ancient morally defined cooperation had on the survival and success of the human species. One might suggest that such a philosophical relational positioning of people may present an anthropocentric view of the world – as the humanity-centred design seems. However, Ramose (1999) argues that in Ubuntu, people and nature are considered interdependent, so care for relations between people also implies care for the natural environment. Likewise, Etieyibo (2017, pp. 633–634) goes to great lengths to also argue that Ubuntu is not anthropocentric, but it "promotes a much better attitude towards the environment or environmental sustainability than the current dominant ethical orientation that is welded to capitalism." It is against this robust co-existence of people with the environment that local indigenous and embedded Afrikan approaches to sustainable design should be built.

The way that education inducts students into engaging with the world around them is critical to changing dominant paradigms (Jansen, 2019). Postcolonial theorist Mbembe (2015) highlights that to set our institutions firmly on the path of future knowledge, we need to reinvent a classroom without walls in which we are all co-learners, a university that is capable of convening various publics in new forms of assemblies that become points of convergence of and platforms for the redistribution of different kinds of knowledge.

Educating design students to appreciate the relational concept of Ubuntu, the acknowledgement of the interdependence of people with each other and the natural world, could begin to change some of the hierarchical and patriarchal pedagogies that have tended to distort significant aspects of design education in the Global North, as well as its influence design education on the Global South. An Ubuntu-inspired approach to design education might parallel Anil Gupta's work in India's similar Global South context.

Agriculturalist turned economist Anil Gupta from the Indian Institute of Management and founder of the Honey Bee Network questions:

> Why is it . . . that the designers of pedagogies and curricula, policies and programmes the world over neglect the need for learning from knowledge rich-economically poor people? Why are there so few papers on innovations by workers in [the] organised and unorganised sector compared to managerial innovations?
>
> *(Gupta, 2012, p. 29)*

In his book *Grassroots Innovation: Minds on the Margin Are Not Marginal Minds*, Gupta (2012) describes how for the last two decades, he has biannually walked a *Shodhyatra* or 'journey on foot' searching for knowledge, creativity, and innovation at the grassroots in India. Thus far, Gupta has covered thousands of kilometres and partnered with various governmental organisations to contribute towards the world's largest open-source innovation platform. The Honey Bee Network (2020) has helped to document and, in partnership with India's National Innovation Foundation (2020), to protect the intellectual property of over 200 000 innovations as part of a grassroots to global strategy for knowledge-based approaches to poverty alleviation and employment generation. This approach to scaling localised indigenous knowledge from the Global South could be highly inspirational for similar Afrikan contexts.

The sustainable design approaches discussed in this chapter use a participatory design approach. The approaches place the design wisdom in the hands of the designer. The designer supposedly knows what a consumer or community requires, particularly in a 'developing' context. However, an Ubuntu-inspired approach to Afrikan sustainable design would expect a horizontal, empathic, caring, authentic relationship between the expert designer and the community lay designers or local experts (Campbell, 2017). A key aspect of Ubuntu is to building interrelationships among the community. Therefore, Afrikan design education should take inspiration from Afrikan indigenous knowledge systems and the passions and experts who have found creative ways to solve their problems, within limited means, in their contexts. In such an approach, designers are tasked with amplifying pre-existing endogenous creative activities or interventions, not to celebrate informality but to acknowledge that the real experts are the lay designers.

Afrikan Sustainable Design Projects

The following section discusses two case studies of Afrikan sustainable design. They are not perfect solutions but are a start towards a more authentic approach to sustainable product design in Afrika, wherein the indigenous philosophy of Ubuntu inspires both cases.

Beegin Bee Bunka Beehive

Beegin started in 2016 as a University of Johannesburg Bachelor of Technology – Industrial Design student research project in South Africa. It emerged out of a broader research project that had been ongoing since 2013 called *iZindaba Zokudla* in isiZulu, which means 'conversations about the food we eat together.' *iZindaba Zokudla* made use of the facilities of the University of Johannesburg, Soweto Campus to bring a wide range of stakeholders together to create opportunities for more sustainable urban agriculture and entrepreneurship in the Johannesburg food system (iZindaba Zokudla, 2022). The inception of *iZindaba Zokudla* and bringing local expert urban farmers together to learn from and connect, creating various forms of social capital (Malan, 2015; Malan & Campbell, 2014), was inherently aligned with the socio-environmental relational conception of Ubuntu.

Through the designer's involvement in the *iZindaba Zokudla* Farmers' Lab (Figure 2.1), Ivan Brown met many emerging urban farmers trying to keep bees but with limited equipment and apiary knowledge. Most of these urban farmers kept bees because they knew having pollinators on their farms increased their crop productivity. Furthermore, honey was a valuable commodity, resulting in almost R3000 ($200) worth of honey per season per hive – at the time, the average monthly household income of most South African households. Therefore, such a valuable crop added significant resilience to emerging and marginalised urban farmers.

Inspired by the passions of these emergent beekeepers, Brown (2017) further explored the wide range of problems beekeepers face in South Africa; these included the loss of beehives due to theft, vandalism, fires, honey badgers, weathering and insect infestation. Using a participatory humanity-centred design process inspired by the concept of appropriate technology (Brown & Campbell, 2017) (Figure 2.2), Brown collaborated with five urban farmers who were either interested in becoming beekeepers or had tried to keep bees in the past. He also worked with six expert beekeepers, who developed, tested and refined a range of Beegin beehives together with the emergent beekeepers.

Using a humanity-centric approach, all the hives were co-designed by the emergent and expert beekeepers to accommodate their needs. However, the

FIGURE 2.1   iZindaba Zokudla Farmers' Lab

Source: iZindaba Zokudla (2022) (Naudé Malan)

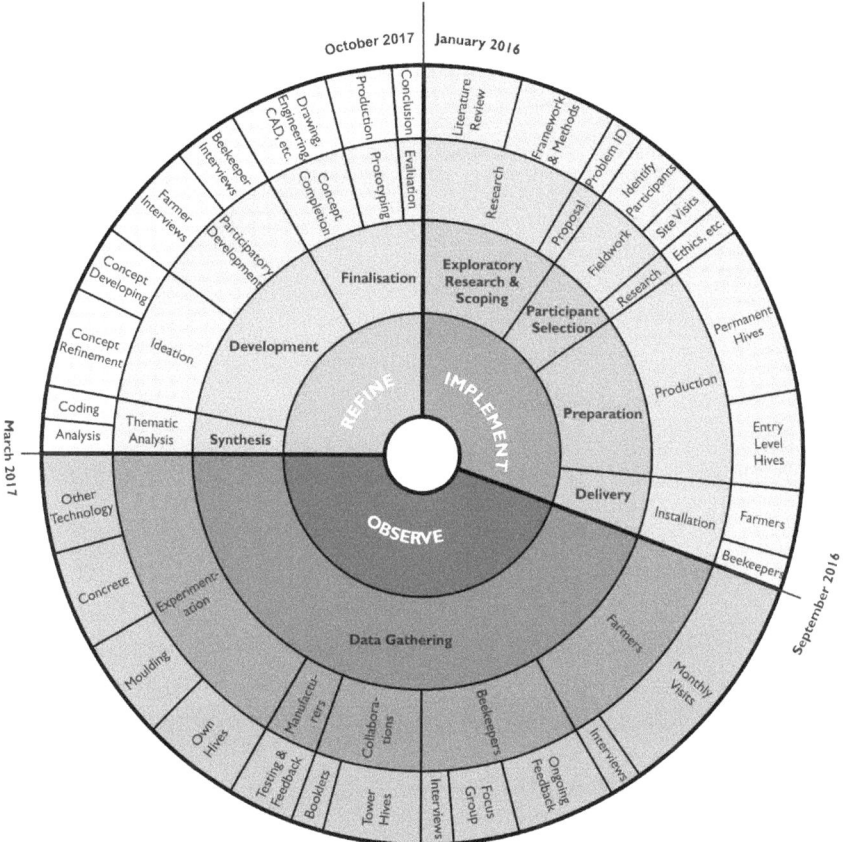

FIGURE 2.2 Beegin design research process

Source: Brown (2017, p. 27)

beehives were designed from the point of view of supporting bee health and natural behaviour, an additional bee-centric approach which was consistent with a socio-ecological conception of Ubuntu. The final design outcome (Figure 2.3), the Beegin Bee Bunka, is a lightweight concrete beehive, a durable, low-cost alternative to wood, and the moulding tools for making them. The design was based on the standard Langstroth hive, with the new Beegin hives protecting both beekeepers and bees from hive losses.

A key innovation in the hive is using lightweight concrete in its manufacture – in field testing, this was found to increase bee productivity by up to 40%. This significant productivity improvement was due to the insulating properties of the lightweight concrete composite hive, which meant the bees needed less energy to heat or cool the hive. The bees were able to spend more time producing honey.

FIGURE 2.3 The final Beegin Bee Bunka beehive

Source: Ivan Brown (2017)

FIGURE 2.4 Urban farmers learning to make Beegin hives with initial prototype moulds

Source: Ivan Brown (2017)

Instead of shipping the giant bulky beehives around the world, which would be costly in terms of financial and environmental impacts, the Beegin business model was conceived on a decentralised model of supply and manufacture. Beegin sells the moulds and production tools in an innovative, open manner for people to make their beehives and potentially begin a local beehive production business (Figure 2.4). Key indicators of the success of this

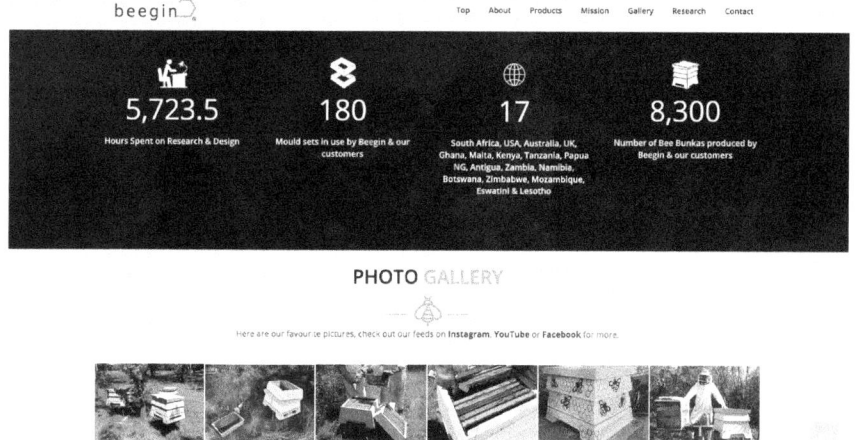

FIGURE 2.5 Beegin website and Bee Bunka impact

Source: Beegin (2022)

approach are the wide range of different experimental fillers various producers have tried and the uptake of moulds and hives around the world. As of March 2022, Beegin has sold 180 mould sets, which have produced 8300 Bee Bunka hives in 17 countries around the world (Figure 2.5).

The relational concepts of Ubuntu within the Beegin project were extensively explored within the concept of critical citizenship (Campbell & Brown, 2018), with a specific focus on power and love (Kahane, 2010). The positive relationships between those that helped conceive the Beegin beehives, the socio-economic benefits of the Bee Bunka's low-impact decentralised production model and its support of bee health and natural behaviour all acknowledge the positive impact a more relational conception of Afrikan sustainable design may have.

The Learning Network on Sustainable Energy Systems (LeNSes)

LeNSes was a collaborative project between universities in the Global North and the Global South, co-funded by Erasmus+ and facilitated by the Learning Network on Sustainability International (2022). The LeNSes project aimed to promote a new generation of design educators and researchers capable of designing sustainable energy systems for all in Botswana, Kenya, South Africa and Uganda, explicitly focusing on Sustainable Development Goal 7: Affordable and Clean Energy.

Although this project received funding from the European Union (EU), it was conducted in Botswana with a range of local participants, experts and researchers to ensure it was cognisant of local culture. The co-design project was undertaken with a primary school and its community in a township in Gaborone, Botswana. Teachers, students, members of the Parents-Teachers Association (PTA), security guards at the school, a local solar energy products entrepreneur, researchers from a local research institution and the Department of Energy Affairs were all participants in the research and co-design activities. The diverse array of participants allowed for the broader social relations of the context to be considered authentically and Ubuntu inspired.

The emergent project brief focused on co-designing a PSS to improve the safety of townships by providing light and security to the public, passages and open spaces which were not lit at night. The project was situated on the school premises, and the various stakeholders contributed their expertise to the project from their local knowledge and experiences (people, technologies and cultures).

The process involved field trips to communities, analysis of the energy context in Botswana, details from project associates to inform the brief, understanding the problem and the stakeholders, generation of ideas, initial concept design, mapping of initial design concepts on a polarity diagram, concept selection and development, field trips to get user feedback, concept and detail design and presentation from student teams to share their work. On its basis of connection, the common good and collaboration (Bremer, 2015), the Ubuntu co-creation process facilitated the generation of results that all stakeholders accepted.

A sample day two plan is shown in Figure 2.6. Through a situated immersive process with local participants, the design students from the University

FIGURE 2.6   Sample day plan

of Botswana could understand some of the energy challenges of the school community. It is worth noting that the proposed solution has moved from a product that can provide light to a PSS – which is aligned with the sustainability design agenda.

Key Results of the LeNSes Project

In order to make the process and the philosophy behind the process available to a broader audience, the project's key results were made available in two formats for the benefit of the design community in Afrika, including design students and design educators. The key results of this project were:

An Open Learning E-package

Following Ubuntu's common good concept, the fruits of the land can be shared with everyone regardless of who owns it. As Bremer (2015) articulated, the common concern of sustainability in Afrika is global. Hence the need to share resources and experiences from a highly interactive and Afrikan design for sustainability projects through an open source, open-use, free-to-download and modified online knowledge repository. The open-source platform contains all project material used and produced during the project, including resource persons and their work in the project. The project resources can be accessed under System Design for Sustainable Energy for All (SD4SEA) at www.lenses.polimi.it. These resources can be used to learn and teach sustainable product service systems applied to renewable and distributed energy systems. Examples in Afrika demonstrate various sustainable energy contexts showing the common good amongst communities and collaboration in the shared use of resources.

Integrated Curricula on Sustainability and
Distributed and Renewable Energy Systems

The knowledge was built into the existing sustainability courses/modules designed at the participating universities in Botswana, South Africa, Kenya and Uganda. This knowledge was built into a Design for Sustainable Development course at the University of Botswana. Topics on sustainable PSS and distributed energy systems were integrated into the course description. An assignment was also done with the group taking the course the following year after the completion of the project to engage the students with the new curricula content areas and expose them to the Ubuntu co-creation process (See Figure 2.7).

The solution-seeking process involved conducting user research to understand the problem space, defining the solution space, exploring solutions and proposing scenarios for a solution. Observations are key in Ubuntu

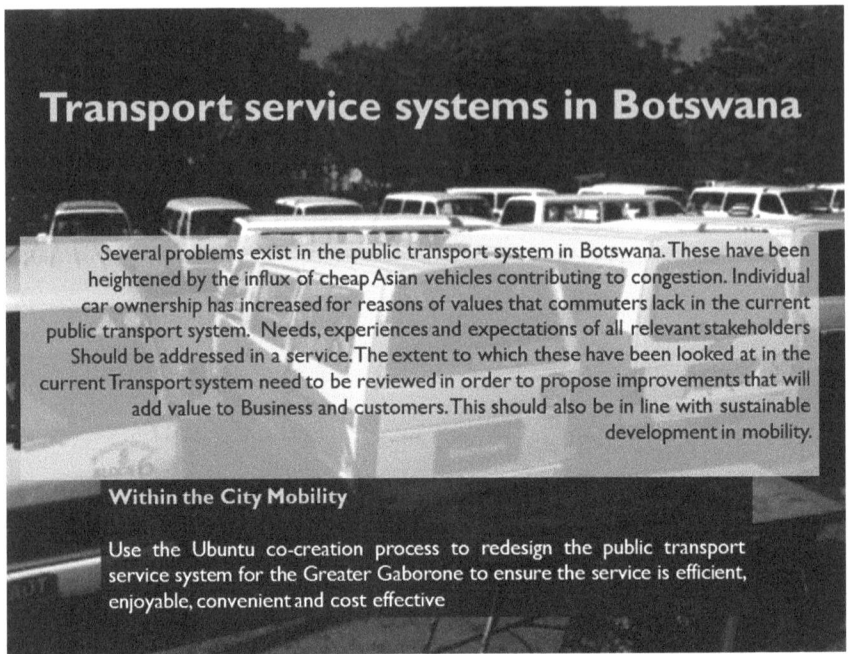

Transport service systems in Botswana

Several problems exist in the public transport system in Botswana. These have been heightened by the influx of cheap Asian vehicles contributing to congestion. Individual car ownership has increased for reasons of values that commuters lack in the current public transport system. Needs, experiences and expectations of all relevant stakeholders Should be addressed in a service. The extent to which these have been looked at in the current Transport system need to be reviewed in order to propose improvements that will add value to Business and customers. This should also be in line with sustainable development in mobility.

Within the City Mobility

Use the Ubuntu co-creation process to redesign the public transport service system for the Greater Gaborone to ensure the service is efficient, enjoyable, convenient and cost effective

FIGURE 2.7   Assignment description for transport service systems in Botswana

co-creation. Non-obtrusive observations formed a more significant part of understanding how users consume the public transport service. In addition to observations, group interviews were used in the context of *lekgotla*, or tribal gathering, to gain insights into public transport issues from the users' point of view and eventually address them. The *lekgotla* approach makes group interviews acceptable and fruitful in Afrikan settings since it relates to tribal gatherings where consultations are made, and a consensus is reached. The outcomes of the assignment were attractive solutions geared towards improving customer experiences and service efficiency, including reducing energy consumption by the service.

Conference Proceedings (Delfino & Vezzoli, 2016)

Activities of the LeNSes project were concluded with a conference in Cape Town, sharing experiences from Botswana, Kenya, Uganda and South Africa. The Ubuntu philosophical co-design approach of designing for the common good was evidenced in work from various Afrikan countries. The research presented showed examples of implementation of renewable energy in Afrikan communities for various applications such as cooking; pumping drinking water; generation of electricity; Afrocentric pedagogical approaches in Botswana, South Africa and Kenya; as well as demonstration projects for social impacts of product service system design and distributed renewable energy systems in rural communities in Kenya.

Designing Sustainable Energy for All (Vezzoli et al., 2018) – A Transdisciplinary Book

The book proposes to enable Afrika to harness and exploit renewable energy abundantly available from various sources in the continent and utilise it for social and economic development. Contributions from the LeNSes project partner institutions contributed to the book, with four of the seven co-authors of the book and five contributors from Afrikan countries. Although the book focuses on worldwide use, the context is Afrikan, referencing policies and strategies in Afrikan countries and work from higher education institutions, local companies and practitioners in the LeNSes Afrikan partner countries. The book has also been published as open access to make it widely accessible free of charge to many people across the globe (https://link.springer.com/book/10.1007/978-3-319-70223-0).

Conclusion

In this chapter, we have explored what an endogenously inspired conception of Afrikan sustainable design might be. Growth-based human development was contextualised versus the limited resources of a finite planet and how this led to a shift in post-war development into its more recent refinement in the Sustainable Development Goals. The chapter then covered the parallel emergence of the discipline of industrial design and its refinement towards a more sustainable approach to design. In alignment with the focus of this book, a more Afrikan conception of sustainability was then explored with a particular focus on the indigenous Afrikan philosophy of Ubuntu – the inseparable relationship between people and the natural environment.

Two design case studies, the LeNSes project in Botswana and Beegin in South Africa, explored Afrikan approaches to sustainable design. Despite different origins, both cases identified elements of the philosophy of Ubuntu, which brought into focus the highly localised interrelation between designers, community and the environment. Although it has been noted that these are not perfect examples, they are a beginning of a more authentic, localised approach to Afrikan design for sustainability that sincerely acknowledges indigenous knowledge and expertise in developing local industrial design products, services and systems. The fact that both cases emerge from design education institutions is a positive move towards a more authentic approach to design education in Afrika. The next generation of designers will be well-equipped to consider sustainability issues from the ethos of Ubuntu. An Afrikan approach to sustainable design moves away from products to services, and the idea of the designer as an expert towards an appreciation of lay designers collaboratively and creatively amplified through local design action.

Reference List

Baha, E., Groenewoud, A., & Mensvoort, K. (2014). Servitisation of products as an approach for design-driven innovation. *Proceedings of the Fourth Service Design and Service Innovation Conference*, 099, pp. 154–163. https://servdes.org/wp/wp-content/uploads/2014/06/Baha-E-Groenewoud-A-van-Mensvoort-K.pdf.

Baines, T., & Lightfoot, H. (2013). *Made to Serve: How Manufacturers Can Compete Through Servitization and Product Service Systems*, 2nd ed. New Jersey: John Wiley & Sons.

Beegin. (2022). *BEEGIN.* https://www.beegin.co.za/.

Bello, D. C., Radulovich, L. P., Javalgi, R. R. G., Scherer, R. F., & Taylor, J. (2016). Performance of professional service firms from emerging markets: Role of innovative services and firm capabilities. *Journal of World Business*, 51(3), pp. 413–424.

Bowles, S., & Gintis, H. (2011). *A Cooperative Species: Human Reciprocity and Its Evolution*. Princton, NJ: Princton University Press.

Bremer, M. (2015, November 5). How to co-create change in an Ubuntu Circle. *Marcellabremer.* www.marcellabremer.com/blog/how-to-co-create-change-in-an-ubuntu-circle/.

Brown, I. L. (2017). *An Appropriate Technology System for Emergent Beekeepers: Field Testing and Development Towards Implementation*. Masters thesis, University of Johannesburg. http://hdl.handle.net/10210/263178.

Brown, I. L., & Campbell, A. D. (2017). Beegin: Redoing beekeeping in Southern Africa by designing for outcomes. In A. L. Bang, M. Mikkelsen, & A. Flinck (Eds.), *Cumulus REDO Conference Proceedings: Design School Kolding 30 May–2 June 2017*, pp. 169–178. Kolding, Denmark: Design School Kolding and Cumulus International Association of Universities and Colleges of Art, Design and Media. https://www.designskolenkolding.dk/en/publications/redo-cumulus-conference-proceedings.

Campbell, A. D. (2013). Designing for development in Africa: A critical exploration of literature and case studies from the disciplines of industrial design and development studies. *Proceedings of the Gaborone International Design Conference (GIDEC) 2013: Design Future: Creativity, Innovation and Development*, University of Botswana, Gaborone.

Campbell, A. D. (2017). Lay designers: Grassroots innovation for appropriate change. *Design Issues*, 33(1). https://doi.org/10.1162/DESI_a_00424.

Campbell, A. D., & Brown, I. L. (2018). A potential difference model for educating critical citizen designers: The case study of the beegin appropriate beekeeping technology system. In E. Costandius & H. Botes (Eds.), *Educating Citizen Designers in Southern Africa*, pp. 85–104. Stellenbosch: African Sun Media. http://www.jstor.org/stable/j.ctv1nzfwm2.1.

Ceschin, F., & Gaziulusoy, İ. (2020). *Design for Sustainability: A Multi-level Framework from Products to Socio-technical Systems*. Routledge. www.taylorfrancis.com/books/oa-mono/10.4324/9780429456510/design-sustainability-fabrizio-ceschin-i%CC%87dil-gaziulusoy.

Cradle to Cradle Products Innovation Institute. (2022). *Home – Cradle to Cradle Products Innovation Institute*. www.c2ccertified.org/.

Crutzen, P. J., & Stoermer, E. F. (2000). The anthropocene. *IGBP Global Change Newsletter*, 41, pp. 17–18.

Delfino, E., & Vezzoli, C. (2016). Sustainable energy for all by design. *Proceedings of the LeNSes Conference*, Edizioni POLI.design, Milan.

De Lille, C. S. H., Roscam Abbing, E., & Kleinsmann, M. S. (2012). A designerly approach to enable organisations to deliver product-service systems. *International DMI Education Conference: Design Thinking: Challenges for Designers, Managers and Organisations*.

Eisenstein, C. (2013). *The More Beautiful World Our Hearts Know is Possible*. Boston: North Atlantic Books.

Etieyibo, E. (2017). Ubuntu and the environment. In A. Afolayan & T. Falola (Eds.), *The Palgrave handbook of African philosophy*, pp. 633–657. New York: Palgrave Macmillan US. https://doi.org/10.1057/978-1-137-59291-0_41.

Fiksel, J. (2006). Sustainability and resilience: Toward a systems approach. *Sustainability: Science, Practice and Policy*, 2(2), pp. 14–21.

Gebauer, H., Joncourt, S., & Saul, C. (2016). Services in product-oriented companies: Past, present, and future. *Universia Business Review*, 49, pp. 32–53.

Goering, L. (2021, November 3). Forget net-zero: Meet the small-nation, carbon-negative club. *Reuters*. www.reuters.com/business/cop/forget-net-zero-meet-small-nation-carbon-negative-club-2021-11-03/.

Gupta, A. K. (2012). Innovation for the poor by the poor. *International Journal of Technological Learning, Innovation and Development*, 5(1/2), pp. 28–39.

Hickel, J. (2020, September 30). The world's sustainable development goals aren't sustainable. *Foreign Policy*. https://foreignpolicy.com/2020/09/30/the-worlds-sustainable-development-goals-arent-sustainable/.

Honey Bee Network. (2020). *Honey Bee Network*. http://honeybee.org/.

IPCC. (2021). *IPCC Climate Report on Climate Change*. www.instagram.com/p/CSwWImgMlgT/?utm_source=ig_web_copy_link.

iZindaba Zokudla. (2022). *About*. IZindaba Zokudla. www.izindabazokudla.com/about.

Jansen, J. D. (Ed.). (2019). *Decolonisation in Universities: The Politics of Knowledge*. Johannesburg: Wits University Press.

Kahane, A. (2010). *Power and Love: A Theory and Practice of Social Change*. Oakland: Berrett-Koehler Publishers, Inc.

Katz, R. (2011). *Boiling Energy: Community Healing Amoung the Kalahari Kung*. Cambridge, MA: Harvard University Press (Original work published 1937).

Kideghesho, J. R. (2008). Co-existence between the traditional societies and wildlife in western Serengeti, Tanzania: Its relevancy in contemporary wildlife conservation efforts. *Biodiversity and Conservation*, 17(8), pp. 1861–1881. https://doi.org/10.1007/s10531-007-9306-z.

Kjaer, L. L., Pigosso, D. C., Niero, M., Bech, N. M., & McAloone, T. C. (2019). Product/service-systems for a circular economy: The route to decoupling economic growth from resource consumption? *Journal of Industrial Ecology*, 23(1), pp. 22–35.

LeNS International. (2022). *Learning Network on Sustainability International*. www.lens-international.org/.

Malan, N. (2015). Design and social innovation for systemic change: Creating social capital for a farmers' market. *Proceedings of the Cumulus Conference, Milano 2015:*

The Virtuous Circle: Design Culture and Experimentation, pp. 919–930, Milan. http://www.d4t.polimi.it/wp-content/uploads/2019/02/cumulus-2015-024-Envisioning.pdf.

Malan, N., & Campbell, A. D. (2014). Design social change and development: A social methodology. *Proceeding of Cumulus Johannesburg: Design with the Other 90%: Changing the World by Design*, pp. 94–101, Johannesburg. https://cumulusassociation.org/wp-content/uploads/2021/10/CumulusJoburgProceedings_.pdf.

Maldonado-Torres, N. (2006). Césaire's gift and the decolonial turn. *Radical Philosophy Review*, 9(2), pp. 111–138.

Manzini, E., & Vezzoli, C. (2003). A strategic design approach to develop sustainable product service systems: Examples taken from the 'environmentally friendly innovation' Italian prize. *Journal of Cleaner Production*, 11(8), pp. 851–857.

Masson-Delmotte, V., Zhai, P., Pirani, A., Connors, S. L., Péan, C., Berger, S., Caud, N., Chen, Y., Goldfarb, L., Gomis, M. I., Huang, M., Leitzell, K., Lonnoy, E., Matthews, J. B. R., Maycock, T. K., Waterfield, T., Yelekçi, Ö., Yu, R., & Zhou, B. (Eds.). (2021). *Climate Change 2021: The Physical Science Basis. Contribution of Working Group I to the Sixth Assessment Report of the Intergovernmental Panel on Climate Change.* Cambridge: Cambridge University Press.

Mbembe, A. (2015). *Decolonising Knowledge and the Question of the Archive.* Africa Is a Country. https://africaisacountry.atavist.com/decolonizing-knowledge-and-the-question-of-the-archive.

McDonough, W., & Braungart, M. (2002). *Cradle to Cradle: Remaking the Way We Make Things.* New York: North Point Press.

Meadows, D., Randers, J., & Meadows, D. (2004). *Limits to Growth: The 30-Year Update.* Vermont: Chelsea Green.

Morelli, N. (2003). Product-service systems, a perspective shift for designers: A case study: The design of a telecentre. *Design Studies*, 24(1), pp. 73–99.

Mugumbate, J. R., & Chereni, A. (2020). Editorial: Now, the theory of Ubuntu has its space in social work. *African Journal of Social Work*, 10(1), Article 1. www.ajol.info/index.php/ajsw/article/view/195112.

National Innovation Foundation – India. (2020). *About NIF | National Innovation Foundation-India.* http://nif.org.in/aboutnif.

Ogude, J. (2019). *Ubuntu and the Reconstitution of Community.* Bloomington: Indiana University Press.

Papanek, V. (1985). *Design for the Real World: Human Ecology and Social Change.* Chicago: Thames & Hudson (Original work published 1971).

Phills, J. A., Deiglmeier, K., & Miller, D. T. (2008). Rediscovering social innovation. *Stanford Social Innovation Review*, 6(4), pp. 34–43.

Porter, M. E., & Kramer, M. R. (2019). Creating shared value. In G. G. Lenssen & N. C. Smith (Eds.), *Managing Sustainable Business: An Executive Education Case and Textbook*, pp. 323–346. Heidelberg: Springer. https://doi.org/10.1007/978-94-024-1144-7_16.

Ramose, M. B. (1999). *African Philosophy through Ubuntu.* Kadoma: Mond Books.

Rapitsenyane, Y., Njeru, S., & Moalosi, R. (2019). Challenges preventing the fashion industry from implementing sustainable product service systems in Botswana and Kenya. In A. Gwilt, A. Payne, & E. A. Ruthschilling (Eds.), *Global Perspectives on Sustainable Fashion*, pp. 236–245. Camden: Bloomsbury Visual Arts.

Rist, G. (1999). *The History of Development: From Western Origins to Global Faith.* Cape Town: University of Cape Town Press.

ServiceDesign.org. (n.d.). *Definition of Service Design*. www.servicedesign.org/glos-sary/service_design/ [Accessed 17 March 2011].

Stare, M., & Križaj, D. (2018). Crossing the frontiers between touch points, innovation and experience design in Tourism. In A. Scupola & L. Fuglsang (Eds.), *Services, Experiences and Innovation*, pp. 81–106. Cheltenham: Edward Elgar Publishing.

Tan, A. R., & McAloone, T. C. (2006). Characteristics of strategies in product/service-system development. In D. Marjanovic (Ed.), *Proceedings of the DESIGN 2006 9th International Conference on Design*, pp. 1435–1442. Dubrovnik, Croatia: Faculty of Mechanical Engineering and Naval Architecture.

Tomiyama, T. (2001). Service engineering to intensify service contents in product life cycles. *Proceedings Second International Symposium on Environmentally Conscious Design and Inverse Manufacturing*, pp. 613–618, Tokyo, Japan. https://doi.org/10.1109/ECODIM.2001.992433.

Truman, H. S. (1949). *Inaugural Address*. www.presidency.ucsb.edu/ws/.

Tukker, A. (2015). Product services for a resource-efficient and circular economy – A review. *Journal of Cleaner Production*, 97, pp. 76–91.

Tutu, D. (2012). *No Future Without Forgiveness*. London: Ebury Publishing.

United Nations. (1987). *Our Common Future, From One Earth to One World [Brundtland Report]*. https://sustainabledevelopment.un.org/content/documents/5987our-common-future.pdf.

United Nations. (2020). *Human Development Report 2020*. http://hdr.undp.org/en/2020-report/download.

United Nations Department of Economic and Social Affairs. (2022). *The 17 Goals*. https://sdgs.un.org/goals.

Vargo, S. L., & Lusch, R. F. (2004). Evolving to a new dominant logic for marketing. *Journal of Marketing*, 68, pp. 1–17.

Vezzoli, C., Ceschin, F., & Diehl, J. C. (2021). Product-service systems development for sustainability. A new understanding. In C. Vezzoli, B. Garcia Parra, & C. Kohtala (Eds.), *Designing Sustainability for All: The Design of Sustainable Product-Service Systems Applied to Distributed Economies*, pp. 1–21. Cham: Springer International Publishing. https://doi.org/10.1007/978-3-030-66300-1_1.

Vezzoli, C., Ceschin, F., Osanjo, L., M'Rithaa, M. K., Moalosi, R., Nakazibwe, V., & Diehl, J. C. (2018). *Designing Sustainable Energy for All: Sustainable Product-Service System Design Applied to Distributed Renewable Energy*. Cham: Springer Nature.

World Design Organization. (2015). *Definition of Industrial Design*. https://wdo.org/about/definition/.

Zondi, S. (2021). A fragmented humanity and monologues: Towards a diversal humanism. In M. Steyn & W. Mpofu (Eds.), *Decolonising the Human: Reflections from Africa on Difference and Oppression*, pp. 224–242. Johannesburg: Wits University Press.

3

NEW PRODUCT DEVELOPMENT PROCESS

A Contextual Approach

Yaone Rapitsenyane and Patrick Sserunjogi

Introduction

New product development (NPD) continues to evolve, intensifying global competition, rapid technology change and shifting patterns in international market opportunities. This has been enhanced by design thinking processes, as they improve the viability and success of new products (Paula et al., 2018). These dynamics continue to compel companies to invest in NPD and stay in business by developing attributes of competitiveness and exploring new offerings or market options.

Accelerated by factors of influence such as functional, symbolic values and economics, product development approaches are forever evolving and highly contextualised. Considering several challenges and other dynamics in product development approaches, economic, societal and individual needs are the most significant emotional factors influencing NPD processes (Ulrich & Eppinger, 2012).

Context is a set of characteristics which describe the situation of a method's application (Kornyshova et al., 2010). Context is an influential notion that fuels the particularisation of meaning and provides coherence (Van Oers, 1998) in a given culture, time or experience. Context, in this section, is divided into three aspects; the situation aspect, the social-cultural dimension and the economic dimension. The situational aspect of context refers to the environment, time and place. The socio-cultural context refers to culture, customs, roles and social status. Finally, the economic contexts refer to investment and consumption, which are core in NPD processes. Product design and development has a solid experiential emphasis, which predisposes it as a suitable platform for exploring NPD through an interdisciplinary approach

DOI: 10.4324/9781003270249-5

(Beard & Wilson, 2006) and integrating strategies that make product development fit the context.

Companies have realised the importance of customising general NPD processes to meet their needs in the idea-to-launch process. Customisation of NPD processes gives organisations greater control in ensuring successful product development. Customisation also allows contextualisation of NPD to ensure that it addresses the market's needs to avoid the failure of super-imposed approaches unrelated to addressing needs and problems in each context (Rapitsenyane, 2014; Ruele, 2015). This approach also makes the NPD meaningful and fruitful as an innovation process, as it represents how products are realised from opportunity spaces to implementation and launch.

This chapter discusses NPD in product design as a complementary process to industrial design and engineering design (Cross, 2000). Within the design and development of a new product, a good balance of engineering innovation, easy-to-use functionality, aesthetics, and safety should be maintained, hence the need for a design-driven NPD process. Kotler and Rath (1984) argue that using design by companies' marketing departments makes it a powerful strategic tool in matching customers' requirements to product-related attributes. Adopting this trend by companies demonstrates that design has taken a leadership position in NPD, which expands beyond traditional design tasks to include direct interface with customers through co-creation and facilitation of customer-to-customer interactions (Taheri et al., 2021). This role addresses the gap between design teams and marketing departments (Bliijlevens & Ranscombe, 2016) and extends to helping organisations envision the future, develop and implement a strategy and cultivate a culture of innovation (Taheri et al., 2021).

Climate change has challenged designers to ensure processes, tools, methods and solutions are geared towards developing sustainable futures. Design contributes enormously to climate change by extracting resources to design and develop products, services and systems. This contribution is mainly attributed to decisions about the product, 70% of which are locked in at the design stage (Bhamra & Lofthouse, 2007; Waage, 2007; Salari & Bhuiyan, 2018). The design stage is where early disassembly considerations, for instance, would have resulted in easy component and material recovery at the end-of-life of the product, hence lower costs. These decisions can be systematically made if considerations are integrated into NPD learning. Therefore, this chapter aims to develop and propose an NPD process appropriate to inform design practice in emerging economies.

New Product Development

New product development refers to a process of product planning and development that extends from the business strategy of the client company to the production, marketing and distribution of a product (Cross, 2000). Ulrich and Eppinger (2012) and Kim et al. (2016) view it as a process organised through a set of activities starting with a need for identification through design, development, production, marketing and selling, and distribution of the solution to end users. Identifying needs or market opportunity spaces is an essential aspect of the process that ensures the value created solves real-life problems. NPD has become an integrated, horizontal and interdisciplinary structure with a high customer focus (Zhu et al., 2019) to aid knowledge sharing across product development teams.

This complexity has enhanced the product development process and has given it a more strategic outlook (Gao & Bernard, 2018; Stock et al., 2021). This approach speaks directly to the business strategies of most companies that position customer satisfaction at the core of their strategy and aligns with findings from Ulrich and Eppinger (2012) that propose an intellectual and organisational application of NPD beyond just the physical production of products. Strategic decisions are intelligent, and so is creativity – the ability to develop novel ideas that promote innovation (Fernandes et al., 2009).

At every phase of the product development process, teams in an organisation undertake different tasks and responsibilities. The generic product development processes show these tasks with an integrated approach at each stage (Ulrich & Eppinger, 2000; Ulrich & Eppinger, 2003; Pugh, 1991; Baxter, 1995). An earlier model by Olsson (1976) still provides the basis of this integrated approach with a five-phase model in which four business areas run parallel to each other, addressing specific issues. These processes differ from one organisation to the other and can be project dependent but still serve the usefulness of quality assurance, coordination, planning, management and improvement (Ulrich & Eppinger, 2000) to the organisation's development efforts.

The process unfolds as a structured flow of activities and information in big companies. This is different in small companies, as the process is informal and unstructured (Capodaglio et al., 2017). In addition to the generic product development process, Ulrich and Eppinger (2012) identified the spiral and complex system development processes. While the spiral process allows many cyclic iterations of design, build and test, the complex system development process enables subsystems to be deconstructed, designed, and tested in parallel. All processes can be triggered by technology-push or market-pull approaches. However, with the rise of human-centred design, it is imperative to adopt human-centred practices to ensure market success for products being developed with users as knowledge-generating agents during the development process (Ollyn, 2015). A spiral process is more aligned with this goal of users as a source of knowledge, as stakeholders can be involved at different stages and contribute feedback that can be used to iterate or improve the primary outcomes. The following section reviews the business context of NPD.

The Business Context of New Product Development

NPD approaches in practice vary between companies. Market-pull enterprises drive their development initiatives based on an identified need and utilise existing technology to solve it (Ulrich & Eppinger, 2012: Lubik et al., 2013). These enterprises are usually user-focused, and product development leads to customised products.

To understand the extent of product-service innovation in manufacturing companies, Tan et al. (2006) discussed two case studies of product manufacturing companies, one building on the product and its technologies and the other building on the user and their experiences. The former relates to a refrigeration component manufacturing company that created a service offering for retailers to detect faults, optimise efficiency and monitor from a centralised site. The latter refers to an office furniture manufacturer who pivoted into a workspace service provider with a specific customer focus. The benefits in both cases have resulted in intimate customer relationships. Both offerings were delivered by establishing a customer-oriented unit separate from the company's product-oriented unit. The customer-oriented units' foundation reflected a different culture and business motivation.

Research highlights the importance of organisations identifying key factors necessary for successfully implementing NPD processes. According to Murthy et al. (2008), formulating and implementing strategies is key to the success of NPD. To illustrate the formulation and implementation of strategy, Cooper (2008) outlines how Emerson Electric, an American company, incorporated a post-launch review into its NPD process. The review allowed the organisation to build continuous improvement, integrating the voice of the customer and making the right project choices. Companies have realised the importance of customising general NPD processes to meet their needs in the idea-to-launch process. Customisation of NPD processes gives organisations greater control in ensuring successful product development. In leading companies, the quest for successful product development has led to a reinvention of innovation processes to ensure they yield results quickly and effectively, reflecting the best practices of user research and creative facilitation approaches (Cooper, 2008). Making NPD processes more flexible by eliminating sequential processes and basing product development on concurrent engineering processes leads to decreased lead times and, ultimately, quicker project launches. Flexibility allows traditionally done activities to overlap and be done in parallel (Cooper, 2008). Flexibility is achieved by adjusting the number of stages in a new product development process depending on the project size (Davis et al., 2014).

Decision-making in NPD processes is vital in ensuring that the entire process is undertaken under strict regulations. According to the Stage-Gate method, a variation of the phase-gate process, gates are where critical decisions are taken regarding the project's progress and nature (Davis et al., 2014). Some decisions must be taken at these gates, including *Go/Kill* decisions, which determine whether a project is viable. If viable, the project will be launched or carried forward, and senior managers will allocate resources.

To avoid having too many gatekeepers, companies have resorted to applying the principle of having 'lean gates.' An example of a lean gate is Johnson & Johnson, as cited by Cooper (2008), where the gate deliverables were cut down to the essentials. The Johnson & Johnson lean gate application works on the premise that gatekeepers attend meetings fully knowledgeable of the project's progress; hence, the forum does not become an educational session. NPD teams and processes have been made multidisciplinary by organisations intending to optimise employee contribution and output while improving the overall process. According to Davis et al. (2014), a multidisciplinary team provides the necessary skills and appropriate representation of organisational functions. A cross-functional team, typically involving R&D, engineering, manufacturing and marketing representatives, has been suggested as the ideal structure for NPD project teams (Davis et al., 2014).

Socio-cultural and Pedagogical Perspectives in New Product Development

This chapter emphasises the value of a socio-cultural context, as a lens of analysis, because of its value as a rich source of ideas, innovation and market insight. As many Afrikan countries are former protectorates or colonies of the West, their socio-cultural values are now based on shared values across Western and local cultures. This gives the local culture a new definition characterised by learning from one another and merging and converging cultural values (Sarala & Vaara, 2010). Fluidity in identity means that successful practices from other existing, new or upcoming contexts can be adopted based on their likelihood of being accepted. For example, values inherent in product and service innovations embody the social contexts that have been developed; thus, their applicability in a different context requires an understanding of socio-cultural values. Developing socio-cultural indicators regarding material, social practices, emotional, technological, and design factors leads to achieving functional satisfaction and pleasure in using products (Moalosi et al., 2005). Although these factors are not exhaustive, they provide a clear picture of the context of producing satisfaction through products and services. These can only be appreciated through the user's involvement on the fuzzy front end and in the design and development stages of the NPD process (Rapitsenyane, 2014).

Adopting NPD processes should borrow from other cultures that do not threaten the central tenets of cultures in each context but rather enhance them. This chapter explores the context of design practice in selected emerging

Afrikan economies. Developing new products in emerging markets is still a challenge to economics and understanding user needs (Srivastava, 2018). As in many other emerging Afrikan economies, design practice still needs to be well established. In such cases, the product import bill is usually very high. For example, in 2017, the total import bill of furniture, metal and metal products, textile and footwear, and wood and paper products in Botswana was BWP 6128.2 million (Statistics Botswana, 2018). In Uganda, the import bill of machines, textiles and wood, among others, amounted to US$2319.33 million in 2017/18 (representing a share of 42.25% of total imports), of which China and India constituted 65.91% (US$1528.62 million) (Uganda Bureau of Statistics, 2013).

These statistics imply a low level of product development activity in many developing Afrikan economies. Sentsho et al. (2007) concluded that Botswana businesses are open to domestic and external competition and are threatened mainly by cheap imports from South Africa and Asia. Regional and global competition cannot be ignored, as it is increasing due to technological changes, freedom of trade and sophistication of communication. Though some of these competitive aspects are prize related, research and development related and skilled human resources related (Jha et al., 2016; Srivastava, 2018), there are non-prize competitive aspects, such as product development tools and approaches, marketing, distribution, after-sales services (Sentsho et al., 2007) and highly price-conscious and demanding customers.

The competitiveness mentioned earlier is a concern for emerging economies and can be addressed through contextualising product development approaches to suit such realities. Congruent with Kahn (2018), NPD is also viewed as an innovation process in which product development techniques and tools are applied at different stages to improve the new product's success. This is a good practice, especially as the innovation process keeps changing. Nowadays, even technology-driven innovation leaders are changing their innovation strategies by opening their innovation processes to external stakeholders (Von Hippel, 2005; Bartl et al., 2010). The innovation process has become more and more of a co-creation process rather than an inward-looking process that relies on the organisation's internal capabilities. A deliberate provision for external stakeholders is necessary for the NPD process, where socio-cultural considerations are made to ensure that the NPD process is more suitable to a given context.

Most of the research and experiences shared on teaching product development are centred on the same philosophical approach of combining theoretical analysis with project work. This usually bridges the gap between theory and practice, using a pedagogical approach called project-based learning (PBL) (Dym et al., 2005; Overbeeke et al., 2004; Moalosi et al., 2012). PBL is one of many instructional models that have led to the compartmentalisation of disciplines in a curriculum, creating confusion among design students who need to learn how to apply theoretical knowledge in their design work (Roozenburg et al., 2008). This is the case in the Delft Design Approach used by TU Delft University through their multidisciplinary thematic courses (Roozenburg et al., 2008). The Delft Design Approach clearly shows the discipline inclination of the themes.

Nevertheless, the translation of knowledge and skills into tangible products and services through a structured and sustainable approach is essential for product designers. Although providing a methodological approach to NPD is valid, it is crucial to reconcile fostering creativity and maintaining discipline (Fernandes et al., 2009). How the process evolves, and feedback is collected and applied from customers and other stakeholders are essential in achieving varying levels of creativity and consistency in product development and its life cycle. The following section examines sustainability issues in NPD.

Sustainability Issues in New Product Development

Sustainable product development (SPD) is a competence that aims to meet human needs and make profits for businesses through a life cycle approach to product designs, with minimal or no environmental impacts, while satisfying other product requirements (Lu et al., 2011). The requirements include functionality, material issues, energy issues, manufacturing requirements, safety concerns, etc. Product design improvements following a life cycle approach clarify the benefits made at one stage and ensure no rebound effect, which may cause the benefits to look insignificant (Klöpffer, 2003).

A life cycle approach, sometimes known as eco-design (Vezzoli et al., 2018), raises sensitivity to the impact of the product development activity in a 'closed-loop' system. It encourages resource efficiency instead of the linear system called the 'materials economy.' This implies that constraints of the product during use and, most importantly, at the end of the product's life are considered during the early design stage. This consideration occurs prior to any commitments to a final artefact, product or specifications. Companies that focus on NPD success are rapidly being driven by global competitiveness, social demands, regulatory demands and product life cycle issues (Ellram et al., 2008), hence efforts to integrate sustainability considerations at each phase of the product development process (Waage, 2007):

For SPD, it is essential first to design the total product life cycle in order to make reuse/recycling activities more visible and controllable, and then to design products appropriately to be embedded in the life cycle (Kimura & Suzuki, 1996 cited in McAloone & Andreasen, 2004). In this case, a life-cycle approach demonstrates benefits at each stage, giving the designer room for more significant influence at the early stages (Bhamra & Lofthouse, 2007). After a dialogue with experts from academic and business circles, Waage (2007) developed an SPD model (Table 3.1). The model offers a structured approach

TABLE 3.1 Product design and sustainability pathways and questions (Waage, 2007)

Design Pathway	Design Questions	Sustainability Questions	Sustainability Pathway
Understand	What is the problem/need/desire?	What is our vision of a sustainable product, material or enterprise?	Establish sustainability context about client and product
Explore	What are potential solutions?	What are the ecological, social and economic implications of the various solutions?	Define sustainability issues Mapping and sustainability analysis
Define and Refine	What is the best solution?	What is the most sustainable solution? (According to a set of ecological and social parameters)	Assess potential pathways forward – a vision of a sustainable solution
Implement	How will we make it?	How can manufacturing, distribution, use, reuse or end-of-life occur sustainably? How will the product's sustainability attributes be assessed over time?	Receive feedback Create and roll out sustainability-oriented product/service Evaluate and (re)assess in terms of sustainability definition and context

through a four-phased design process in which sustainability is taken up at each phase. This incorporation of sustainability principles early in the design process reinforces the importance of focusing on the 'new front end' or early stages of the product development process as critical to creating product differentiation and gaining a competitive advantage (Cagan et al., 2001).

The following section advances that the NPD should not be treated as a linear practice but as flexible, iterative and adaptable to various development activities. Some case studies will be discussed to illustrate the relevance and usefulness of the NPD process.

Creative Intelligence for Inquiry

The notion of Ubuntu believes that humans possess creative minds, which creative minds link creative thought and actions to the intelligent use of the environment (Jónsdóttir & Gunnarsdóttir, 2017). The intelligent use of the creative powers of human minds leads to the production of ideas and objects that present novel solutions for the betterment of human development. Creative intelligence is the ability to create novel and exciting ideas (Moller, 2005) that culminate in objects of intellectual inquiry (Levin, 1972). This inquiry stems from human creative impulses, expressed through progressive thoughts and actions under creative processes in a purely physical domain (ibid.). Within the domain of NPD, creative intelligence inquiry plays a pivotal role in inspiring human creativity to transform human conditions into broader developmental contexts. Further, it supports designers in looking for possibilities that may not have been explored.

Transferring this capability in the business context means focusing on the product and the entire innovation process, especially early stages where identifying business opportunities as an entrepreneur, is critical. Even though business tools are still needed for this process, a design attitude is necessary to provide a unique problem-solving mindset. It empowers designers to think of practical solutions from a human-centred design point of view. This approach is supported by Boland and Collopy (2004), who found that a design attitude can drive profitability and human satisfaction.

Creative intelligence has led to the emergence of various design approaches, such as the Delft Design Approach and the IDEO design process, among others, with the intent to harmonise design with business and foster sustainability in new product development processes. The Delft Design Approach focuses on understanding the context of use. It adopts an emphatic approach to understanding users and how to derive insights to inspire the design process. In the Delft Design Approach, this attribute is characterised by understanding and defining. The IDEO human-centred design process addresses this attribute in the observation phase. The end-user is observed to learn and be open to creative possibilities by identifying patterns of behaviour, pain points and difficulties users experience in undertaking specific tasks. A user immersion experience is critical in the IDEO approach. The core competencies supporting creative intelligence include empathy building, immersion experience, user focus and perspective, social and cultural awareness, sustainability considerations and market orientation.

Applying the design approaches to business requires an initial planning phase which mainly comprises a need and market opportunity, identification, justification and primary research. During this phase, a holistic understanding of the market (users, stakeholders, existing offerings and competitors) is vital to define a niche opportunity space in which development decisions are grounded. Under this attribute, sustainability questions relating to the overall solution, material composition and offering type should be in focus. Figure 3.1 shows an example of a baby incubator designed through an emphatic research insight.

Premature babies account for 0.7% of all hospital admissions in Uganda (Egesa et al., 2020). Stakeholders and health workers must scale up

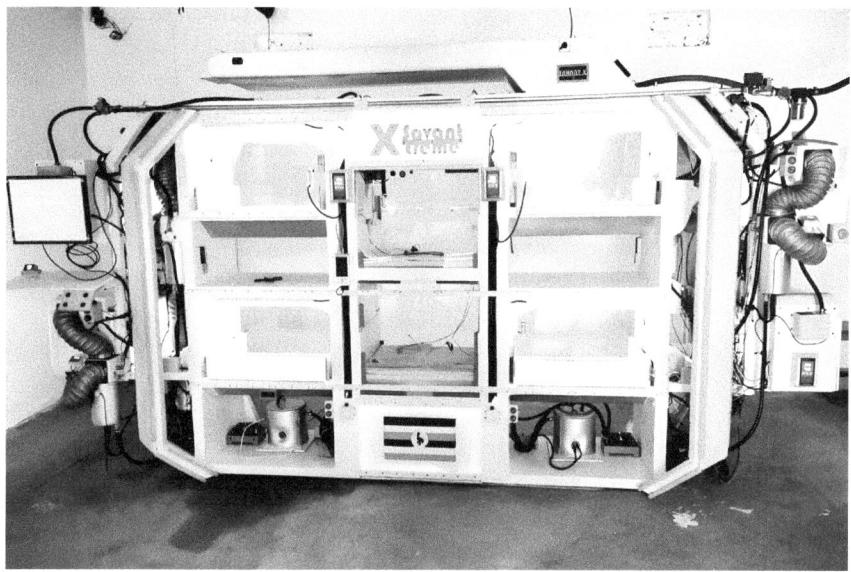

FIGURE 3.1   Baby incubator: Nsamba Christopher, African Space Research Program (2020)

endeavours to address premature birth and enhance the care of premature babies if Uganda is to attain the Sustainable Development Goals target (SDG 3.2) (ibid.). This situation warranted an intervention requiring creative intelligence to develop the largest and most technologically advanced baby incubator that can carry ten babies simultaneously (Figure 3.1).

Creative Engagement for Exploration of Possible Solutions

The define phase comprises development activities that explore possible solutions and culminates in defining the best solution. Co-creation approaches should be used to engage potential users and stakeholders in the development process to benefit the development of a product or service (Hughes, 2014). Co-creation is not a new phenomenon; it has existed in Afrikan communities as one of the tenets of the philosophy of Ubuntu. The philosophy calls for interdependence, inclusivity and inter-subjectivity (Bolden, 2014). The Ubuntu processes empower the community and recognise that local people are a source of experiential knowledge and insight that the design process can leverage to remain contextually relevant and co-create appropriate innovations (Moalosi et al., 2017).

The approach brings possible users' solutions closer to meeting their needs in the best possible way and provides a foundation of ownership of the solution. This provides stakeholders with a platform to help diffuse win-win scenarios across the product life cycle. A life cycle approach allows actors to define their roles and contribute at the right life cycle stage, creating a closed-loop development cycle. Figure 3.2 shows an example where the community co-created a solution to the rise of environmental degradation caused by lit-tered plastic bottles.

The Ubuntu philosophy was applied during the creative engagement when the artists, together with the local communities who collected littered plastic bottles in their communities at Kawempe in Uganda, co-created "The hand that speaks." This approach appreciated all stakeholders in Kawempe as essential ingredients in the product development process to participate and co-create the solution to raise environmental degradation awareness. This project brought possible solutions for transforming littered plastic bottles found in communities for reuse into a product that benefits and raises awareness in the community. The following section develops and proposes an NPD process/model suitable to inform design practice after critically assessing product development processes in Uganda and Botswana.

FIGURE 3.2 "The hand that speaks" Eco Art/RUGANZU BRUNO

The Local Landscape – A Contextual Approach

Small and medium-sized enterprises (SMEs) are increasingly recognised as the lifeblood of modern economies because they contribute significantly to the gross domestic product (Ghobadian & Gallear, 1996). Due to local and global competitive pressure, SMEs must develop products that harmonise with users' experiences and needs. This requires creative intelligence in the process of new product development. However, SMEs in emerging economies face the challenge of developing a contextualised NPD process which they can use to innovate their products and services.

The NPD in Uganda and Botswana are in their infancy and very informal. These two countries' product development and production processes need to be better defined. There is little information on how this process evolves except for sharing ideas at the beginning of every project. While these processes may be documented or available on internal company documents, their capabilities are limited. The intellectual approach behind the production of product concepts is affected by internet access, and sometimes there is a deficiency of understanding on the significance of these processes by SME management.

Despite a lack of information on the processes, the products are still being made to some acceptable level of quality. For example, Figure 3.3 shows quality interiors and furniture designed by Sasa Interiors in Botswana.

Furthermore, Figure 3.4 shows furniture inspired by the local culture (cultural patterns and materials) designed by Mabeo Furniture in Botswana. This

FIGURE 3.3   Sample products from one interior design company

Source: Sasa Interiors, 2020

is one of the few furniture design companies that has managed to break into the global export market as the products display a high level of quality and showcase an international appeal (Figure 3.4).

However, most SMEs need a transparent process. There is no evidence that the products result from a novel design and creative thinking process. Such products may be an imitation or copies of existing products. Remarkably, the line between inspiration and imitation has also gotten blurry unless considered in a legal context (The Fashion Law, 2018; Brown, 2014). This shows a need for more explicit and documented design processes.

In a study conducted by Löfqvist (2009), it was found that in multiple case studies of three established small companies in Sweden, where the relative novelty of the product and the process is low, a linear process was used. In contrast, a cyclic, experimental, and knowledge-creating process was associated with high novelty. With the low relative novelty of product design and the absence of a documented strategy, there are high chances of inspiration

FIGURE 3.4   Mabeo Furniture products at the 2017 Milan Design Week

Source: Mabeo Furniture

closer to imitation in emerging economies. This is notwithstanding that local SMEs still do not have proper design qualifications and resources to develop methodological NPD processes (Rapitsenyane, 2014).

A Proposed NPD Process/Model

This section aims to develop and propose an NPD process/model suitable to inform design practice after critically assessing product development processes in selected Afrikan countries. The process is based on lessons learnt from the literature, case studies and practice experiences of emerging economies design companies on new product development and innovation processes. Through this model, a proposal on approaching new product development in Botswana and Uganda is made, carefully adopting similarities from models used in other contexts. A focus on how to be context-sensitive is captured early in the model so that the innovation process is responsive to market needs at the early stages of the development process.

The proposed NPD process considers the socio-cultural context, sustainability considerations throughout the development process and competencies needed in the model, and reflects on the design practice landscape. These considerations provide a holistic picture of approaching product development and can thus be leveraged by SMEs and higher education design schools to promote professional design practice methodologies in emerging economies.

This model may be the first conceptual model developed for promoting design practice in emerging African economies. It is critically necessary to consider the global economy and competition developments using knowledge (knowledge economy) and sustainability (circular economy). Thus, the model can be customised for specific needs and applications per companies' business demands or requirements (Stock et al., 2021; Gao & Bernard, 2018).

Making the model with fewer stages provides room for adjusting the number of activities within each stage per the demands of the project at hand (Davis et al., 2014). The activities and flow of information throughout the model follow a cyclic approach, as Ulrich and Eppinger (2012) discussed, to allow flexibility, feedback loops and iterations during the development process. A cyclic approach also provides for overlapping activities and, in some cases, running activities concurrently instead of finishing one activity first, as is the point in linear models (Cooper, 2008). The structure of the model is that of a creative process of looking for opportunities and defining the unique opportunity spaces, exploring solutions and validating them before rolling them out into the market. Therefore, the stages of the model have been organised as Inquire, Define, Design, Validate and Deliver (Figure 3.5).

In projects where designers need to work on community projects, there is a need to conduct a scoping exercise. This stage is critical and often needs to be included in the NPD process, especially in emerging economies.

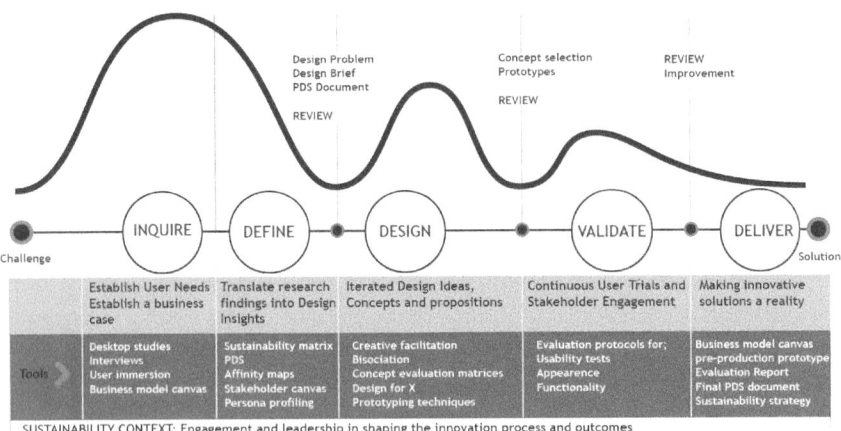

FIGURE 3.5   New product development process

Scoping

This stage involves taking preliminary visits to gain access to the stakehold-ers, develop rapport and build trust and empathy with the community. Trust is fundamental in such societies because any intervention from outside might be considered suspicious. Communities are fatigued as they have been used as research instruments, but they never see the benefits of such research (Moalosi et al., 2017). Therefore, they have started to resist such exploita-tion. In most cases, entering the community from the outside involves getting a gatekeeper (a person who knows the community) to introduce the design team. It also involves meeting the village elders or chief to explain the project objectives and outcomes. During these visits, for the design team to gain the community's trust, they should show humility and respect for their ancestors, traditions, ancient teachings and practices and the environment, and show interest and enthusiasm in learning the local culture. This is a critical stage that is not found in other design processes.

Inquire

The Inquire stage is the research phase of the process/model. After gaining the community's trust and willingness to participate in the project, the design team begins by understanding the community's actual needs. Understand-ing the socio-economic need is vital at this stage to ensure product success; the solution is fit for the context and the market. First, there is a need to get management commitment, which is essential to initiate the design pro-cess. Employee involvement is another critical factor in the first stage of NPD. This stage is ambiguous and chaotic, and many activities are taking place to inform and inspire the exploration of open-ended questions that describe the problem statement. It is often unknown whether the result will be a product or a service. The goal is to define the fundamental problems and opportunities and determine what should not be designed. After that, an outward process is needed that involves user research through an empathic approach to understand user needs from their perspective, market research to identify gaps regarding the competitor's products, market share, environmental and social impacts of competitors and the business case for product development. The designer should identify the product's use and its users in a sustainability context. The possible tools and methods for this stage include literature review, user immersion, empathy interviews, observations and other business tools, such as the business model canvas.

The community and stakeholders identify the problem collectively by sharing their indigenous knowledge, life experiences and social needs. It is essential to understand the problem context and collect information from the community about their customs, norms, beliefs, values and spirituality. The community's social needs become the design team's needs. Their joys and sorrows become the design team's joys and sorrows. It is about putting the problems, interests and circumstances at a higher level of attention. For example, Ubuntu advocates that we are human through our interaction with others. At the end of this phase, the target market needs and an ill-defined problem should be captured in a well-organised research report.

Define

The Define stage offers an opportunity for the design team to gain a deeper understanding and insights from the Inquire stage to frame the problem into a significant design challenge. In this space, the non-technical language of user needs, requirements and other insights should be interpreted, analysed, synthesised and decoded into the technical design language. User research should be conducted based on the users' context and technology, interpret the findings, and find meaning. The design team should have empathy, which is the ability to successfully enter the emotional situation of the community or stakeholders, to listen and feel genuine sympathy because they hear and feel what others share with them from a deeper perspective. When the practice of empathy is cultivated, it deepens the design team and gives them access to more humanness with which they can help others. The information collected, usually through narratives, is "analysed with the community or with the people who understand and can interpret the local people's language, idioms, and proverbs" (Chilisa et al., 2016, p. 357). The team should then frame opportunities and set the design challenge.

More parameters for the project should also be defined at this stage, such as stakeholders, through various stakeholder mapping techniques. Users must also be profiled to allow downstream design activities to refer to these profiles. Since a sustainability scan should have been carried out during the inquiry stage, the requirements should be defined at this stage. These should form part of the product design specifications as early proposals for defining design requirements. The business case should also be updated to reflect inputs from this stage. Possible tools and methods include an innovation matrix, affinity maps, stakeholder canvas, business model canvas, persona templates and total design elements of product design specifications. At the end of the Define stage, the design problem, market opportunity, design brief and product design specifications should be presented to the design team for review. If there is a need for further clarity, the evidence presented should be interrogated again, with a possibility of going back to the Inquire stage.

Design

This stage involves a collaboration between designers and stakeholders in sharing their initial thoughts or ideas about the problem identified. In the Ubuntu philosophy, it is customary to share generously with others. By sharing, one expresses their humanness and finds joy within themselves. The ideation team should be composed of a mixed group of participants: community, stakeholders and designers. The main activity is brainstorming and making random connections in exploring the problem in all facets. This is achieved through solidarity, that is, accomplishing complex tasks collectively. At this stage, personal interests are less important than community needs. Solidarity is necessary for the design team cohesion and commitment, which cultivates a collective mindset. The '*I*' is eliminated, and the '*we*' state of mind becomes predominant in proposing design ideas.

The design team collaborates to transform some ideas into concepts to solve the identified need. Through the teams' creativeness, knowledge sharing and constructive activities, they will reach a consensus on the best solution that satisfies the community based on their needs and context. The design team creates a shared vision through personal understanding and caring for each other. High levels of respect and dignity can lead to high levels of mutual trust, leading to the design team's high performance in designing innovative cultural-sensitive concepts. Shared vision, caring, consensus decision-making, trust, respect and dignity are some of the tenets of the philosophy of Ubuntu that should be displayed in this stage. The team develops rudimentary working models or prototypes to explore functional and structural aspects of a product, service, system or user experience. The prototype is commonly built for demonstration purposes, visualising and testing the design and then modifying it, if necessary, until an acceptable prototype is finally achieved.

Several sustainability pathways proposed during the Define stage should be assessed at this stage, alongside other product considerations as solutions are explored. Making sustainability the context in which solutions are explored ensures a sustainable solution composed of good design characteristics. Ideas should be rapidly sketched and evaluated against the criteria to guide the selection process. Other ideas are discarded and merged through discussion and review to develop the final design concepts. As part of the iterative design approach, the evaluation of initial ideas (2D sketches) should involve stakeholders before any 3D computer-aided design (CAD) work and physical prototypes are developed. Early stakeholders' input ensures their buy-in and provides non-technical input into the solution that reflects their contextual and market dynamics (Ulrich & Eppinger, 2012). Critical design activities include generating ideas through sketching and low-fidelity prototypes, developing concepts and refining the solution through 3D CAD and high-fidelity prototyping and detail design. Prototyping is integral to this stage since stakeholders should collect continuous feedback to contribute to the next iteration.

Once the design team has secured the buy-in of stakeholders, a detailed design should consider costing and technical and engineering drawings to aid the development of the final appearance functional prototype. At this point, stakeholders will be co-owners of the solution since they will have been co-developers from the Inquire stage. Possible tools and methods include creative facilitation sessions, brainstorming, bisociation, concept evaluation, design for X, product costing and testing protocols for the prototype. The end of this stage should be epitomised by alternative product concepts chosen, concept prototypes, engineering drawings and production plans.

Validate

To ensure that the solution meets user needs and that there is a business case for it, in addition to meeting technical requirements, the solution should be validated throughout the development process. At each stage of the development process, there is a criterion which the solution is evaluated against to inform the next steps. The purpose of the evaluation is usually twofold. Firstly, to prove the design in terms of achieving what it was hoped to accomplish regarding function and other performance criteria such as speed, force, accuracy and usability issues such as ease of use, comfort and pleasure. Secondly, designs are usually evaluated to improve them by using feedback from user trials or expert evaluation sessions to identify strengths and weaknesses. A range of prototypes allows for exploring various issues ranging from appearance, function, usability, etc. The prototype evaluation protocols should be developed with selected representative users, who will carry out representative tasks within a typical environment.

The design validation can be effectively done using the Pugh-controlled convergence method (Seperamaniam et al., 2017; Frey et al., 2009). The Pugh matrices can provide significant benefits by improving alternatives and facilitating convergence (Frey et al., 2009). The method provides organised ways of iterated evaluation cycles, refined ideation, and further investigations to address loose ends until the final design is signed off. Effective validation of the prototypes can be achieved by surveys (for example, Likert-item survey), interviews, demonstrations, a single evaluation question often to gain an overall opinion about the product and ranking prototypes evaluated from best to worst (Choi & Sprigle, 2011). The evaluation includes function, performance, sustainability, usability and other performance criteria. Validation is a co-design activity involving the design team and stakeholders through user trials and tests, following evaluation protocols for various design aspects.

Deliver

The Deliver stage is the pre-production stage of the model. Feedback from the design stage is used to develop a fully-fledged functional appearance prototype, which looks like and works like the final product. Where possible, this prototype is made from actual materials and production processes close to those which will produce the final product. This stage provides stakeholders with a proof of concept. Further testing of the prototype is performed in the real market. Sustainable business viability analysis is also done to check if the sustainability criteria have been met and identify ways of making improvements. The business model is also updated and subjected to testing through a scaled-down implementation approach. The ecosystem of the business should be the premise against which the business model is tested so that alterations can be made before the production is scaled up. At the end of this stage, product production documentation, pre-production prototype, testing and evaluation report, updated and final detail design drawings, revised and final product design specifications, a business ecosystem map, updated and final business model canvas and a sustainability-oriented product strategy should be ready to be rolled out. An evaluation criterion for monitoring sustainability targets should also be prepared to ensure that the product life cycle is kept in check as it gains access to the market.

Exit Strategy

In the case of community projects, the design team should devise an exit strategy to ensure the project's sustenance. Often, communities complain that they have been involved in projects that have not empowered them. Such projects usually collapse when the project initiators leave. The exit strategy should be gradual to ensure the community assumes ownership and responsibility for the project and is empowered to sustain the project independently.

Conclusion

This chapter has proposed a new product development model centred on contextualisation techniques concerning aspects that make it a success in emerging economies. The NPD structure comprises various organisational, research, design, prototyping and product-selling techniques. These techniques should be reflected in the way activities are organised. The motivation for this chapter is to provide insight into a space where knowledge is disorganised and undocumented, making it generally limited. To this date, an NPD model has yet to be developed with the ethos of Ubuntu in Botswana and Uganda for use by SMEs or companies. Design practitioners in Afrika may use this spiral model to guide their innovative capabilities. This could alleviate copying problems and limited understanding of design (Rapitsenyane & Bharma, 2013; Rapitsenyane, 2014), as activities at each stage guide the development of a new product.

The developed NPD model can be used for teaching at tertiary institutions, as it allows for several iterations of designing and prototyping until the final solution is achieved. This is a good way of teaching iterative design and prototyping for undergraduate programmes. There is a need for experiential learning to unfold in responding to contextual issues to demonstrate the value of NPD in the emerging context and its economic value to students and their stakeholders. Achieving a synergised NPD process responsive to context-specific needs should be explored, tested and institutionalised.

While this chapter may be limited by its exploratory nature, it is sufficient to stimulate dialogue on organising product development activities. It also opens new horizons for supporting the creative process by encouraging industries from emerging economies to search for design cues in the local culture, practices, ornaments, experiences and objects with valuable meanings and rich with inspiration. As emerging economies in Afrika explore export opportunities, this model allows local companies to finally explore unique offerings to export, which are representative of the local feel and look in the global context. Global competition has become knowledge-intensive and requires sources of this knowledge to make it accessible.

Reference List

Bartl, M., Jawecki, G., & Wiegandt, P. (2010). *Co-Creation in New Product Development: Conceptual Framework and Application in the Automotive Industry*. file:///C:/Users/moalosi/Downloads/Co-Creation_in_New_Product_Development_Conceptual_. pdf https://www.researchgate.net/profile/Michael-Bartl-2/publication/228905562_ Co-Creation_in_New_Product_Development_Conceptual_Framework_and_ Application_in_the_Automotive_Industry/links/00b495327f843567e0000000/ Co-Creation-in-New-Product-Development-Conceptual-Framework-and-Application-in-the-Automotive-Industry.pdf

Baxter, M. (1995). *Product Design, Practical Methods for the Systematic Development of New Products*. London: Chapman & Hall.

Beard, C., & Wilson, J. P. (2006). *Experiential Learning – A Best Practice Handbook for Educators and Trainers*. London: Kogan Page.

Bhamra, T., & Lofthouse, V. (2007). *Design for Sustainability: A Practical Approach*. Hampshire: Gower Publishing, Ltd.

Blijlevens, J., & Ranscombe, C. (2016). Bridging the gap between marketing strategy and design teams: A method to facilitate strategic styling decision-making within a company. *Journal of Design, Business & Society*, 2(2), pp. 217–233.

Boland, R. J., & Collopy, F. (2004). *Managing as Designing*. Stanford, CA: Stanford University Press.

Bolden, R. (2014). Ubuntu. In D. Coghlan & M. Brydon-Miller (Eds.), *Encyclopedia of Action Research*. London: Sage Publications.

Brown, E. (2014). *Inspiration vs Imitation: Where to Draw the Line*? www.designmantic.com/blog/inspiration-vs-imitation/ [Accessed 4 November 2020].

Cagan, J. M., Vogel, C. M., & Nussbaum, B. (2001). *Creating Breakthrough Products: Innovation from Product Planning to Program Approval*. New Jersey: FT Press.

Capodaglio, A., Iacoviello, G., & Neri, G. (2017). Family business: From an informally managed and unstructured model to a structured, formally managed larger enterprise. *Corporate Ownership & Control*, 15(1), pp. 123–132. https://doi.org/10.22495/cocv15i1art12.

Chilisa, B., Major, T. E., Gaotlhobogwe, M., & Mokgolodi, H. (2016). Decolonising and indigenising evaluation practice in Africa: Towards African relational evaluation approaches. *Canadian Journal of Program Evaluation*, 30(3), pp. 347–362.

Choi, Y. M., & Sprigle, S. H. (2011). Approaches for evaluating the usability of assistive technology product prototypes. *Assistive Technology®*, 23(1), pp. 36–41.

Cooper, R. G. (2008). What leading companies are doing to re-invent their NPD processes. *PDMA Visions Magazine*, pp. 6–10.

Cross, N. (2000). *Engineering Design Methods: Strategies for Product Design*. Chichester: John Wiley & Sons, LTD.

Davis, D., Chelliah, J., & Minter, S. (2014). New product development processes in the Australian FMCG industry. *Contemporary Management Research*, 10(1), pp. 3–22.

Dym, C. L., Agogino, A. M., Eris, O., Frey, D. D. and Leifer, L. J. (2005). Engineering design thinking, teaching, and learning. *Journal of Engineering Education*, 94(1), pp. 103–120.

Egesa, W. I., Odong, R. J., Kalubi, P., Yamile, E. A. O., Atwine, D., Turyasiima, M., . . . & Ssebuufu, R. (2020). Preterm neonatal mortality and its determinants at a Tertiary Hospital in Western Uganda: A prospective cohort study. *Pediatric Health, Medicine and Therapeutics*, 11, p. 409. http://doi.org/10.2147/PHMT.S266675.

Ellram, L. M., Tate, W. L., & Carter, C. R. (2007). Product-process-supply chain: An integrative approach to three-dimensional concurrent engineering. *International Journal of Physical Distribution and Logistics Management*, 37(4), pp. 305–330.

The Fashion Law (2018). *When Is Inspiration Just Inspiration and NOT "Imitation"?* www.thefashionlaw.com/when-is-inspiration-just-inspiration-and-not-imitation/ [Accessed 4 November 2020].

Fernandes, A. A., da Silva Vieira, S., Medeiros, A. P. and Natal Jorge, R. M. (2009). Structured methods of new product development and creativity management: A teaching experience. *Creativity and Innovation Management*, 18(3), pp. 160–175.

Frey, D. D., Herder, P. M., Wijnia, Y., Subrahmanian, E., Katsikopoulos, K., & Clausing, D. P. (2009). The Pugh controlled convergence method: Model-based evaluation and implications for design theory. *Research in Engineering Design*, 20(1), pp. 41–58.

Gao, J., & Bernard, A. (2018). An overview of knowledge sharing in new product development. *The International Journal of Advanced Manufacturing Technology*, 94(5–8), pp. 1545–1550.

Ghobadian, A., & Gallear, D. N. (1996). Total quality management in SMEs. *Omega*, 24(1), pp. 83–106.

Gudlavalleti, S., Gupta, S., & Narayanan, A. (2013). *Developing Winning Products for Emerging Markets*. www.mckinsey.com/business-functions/operations/our-insights/developing-winning-products-for-emerging-markets.

Hughes, T. (2014). Co-creation: Moving towards a framework for creating innovation in the Triple Helix. *Prometheus*, 32(4), pp. 337–350.

Jha, S. K., Parulkar, I., Krishnan, R. T., & Dhanaraj, C. (2016). *Developing New Products in Emerging Markets*. http://sloanreview.mit.edu/article/developing-new-products-in-emerging-markets/.

Jónsdóttir, S. R., & Gunnarsdóttir, R. (2017). Creative intelligence for intelligent creations. In *The Road to Independence*, pp. 35–44. Rotterdam: SensePublishers.

Kahn, K. B. (2018). Understanding innovation. *Business Horizons*, 61(3), pp. 453–460.

Kim, Y.-H., Park, S.-W., & Sawng, Y.-W. (2016), Improving new product development (NPD) process by analysing failure cases. *Asia Pacific Journal of Innovation and Entrepreneurship*, 10(1), pp. 134–150. https://doi.org/10.1108/APJIE-12-2016-002.

Kimura, F., & Suzuki, H. (1996). Design of right quality products for total life cycle support. *Proceedings of 3rd International Seminar on Life Cycle Engineering*, Eco-Performance '96, pp. 127–133, Zürich, Switzerland.

Klöpffer, W. (2003). Life-cycle-based methods for sustainable product development. *The International Journal of Life Cycle Assessment*, 8(3), pp. 157–159.

Kornyshova, E., R Deneckère, R. & Claudepierre, B. (2010). Contextualization of method components. Fourth International Conference on Research Challenges in Information Science, pp. 235–246, IEEE, Nice.OI: 10.1109/RCIS.2010.5507383.

Kotler, P., & Rath, G. A. (1984). Design: A powerful but neglected strategic tool. *Journal of Business Strategy*, 5(2), pp. 16–21. https://doi.org/10.1108/eb039054.

Levine, P. H. (1972). Transcendental meditation and the science of creative intelligence. *The Phi Delta Kappan*, 54(4), pp. 231–235.

Löfqvist, L. (2009). Design processes and novelty in small companies: A multiple case study. *Proceedings of the 17th International Conference on Engineering Design*, ICED'09, pp. 265–278, 24–27 August 2009, Stanford University, Stanford, CA. http://urn.kb.se/resolve?urn=urn:nbn:se:kth:diva-25331.

Lu, B., Zhang, J., Xue, D., & Gu, P. (2011). Systematic life-cycle design for sustainable product development. *Concurrent Engineering*, 19(4), pp. 307–324.

Lubik, S., Lim, S., Platts, K., & Minshall, T. (2013). Market-pull and technology-push in manufacturing start-ups in emerging industries. *Journal of Manufacturing Technology Management*, 24(1), pp. 10–27. https://doi.org/10.1108/17410381311287463.

McAloone, T. C., & Andreasen, M. M. (2004). Design for utility, sustainability and social virtues, developing product service systems. *Proceedings of International Design Conference*, Design-04, Dubrovnik, Croatia.

McAloone, T. C., & Andreasen, M. M. (2006). What happens to integrated product development models with product/service-system approaches? *Proceedings of the*

6th Workshop on Integrated Product Development, September 2006, Design Society, Magdeburg.

Moalosi, R., Marope, O., & Setlhatlhanyo, K. N. (2017). Decolonising Botswana's design education curriculum by infusing indigenous knowledge: Botho co-creation process. In M. T. Gumbo & V. Msila (Eds.), *African Voices on Indigenisation of the Curriculum: Insights from Practice*, pp. 66–96. Wandsbeck: Reach Publishers.

Moalosi, R., Molokwane, S., & Mothibedi, G. (2012). Using a design-orientated project to attain graduate attributes. *Design and Technology Education*, 17(1), pp. 30–43.

Moalosi, R., Popovic, V., & Hickling-Hudson, A. (2005). Integration of culture within Botswana product design. In J. Redmon, D. Durling, & A. de Bono (Eds.), *Futureground: Volume 2: Proceedings*, pp. 1–11. Melbourne, Australia: Monash University.

Moller, C. (2005). *Creative Intelligence*. https://clausmoller.com/wp-content/uploads/2020/11/Creative_Intelligence-CMC.pdf.

Murthy, D. P., Rausand, M., & Østerås, T. (2008). *Product Reliability: Specification and Performance*. London: Springer Science & Business Media.

Ollyn, M. G. (2015). *Investigation of User-Centred Approaches to Design Practice in Botswana*. Doctoral thesis, Leicestershire: Loughborough University.

Olsson, F. (1976). *Systematic Design: A Study with the Aim of Systematizing Content and Methods in Connection with Product Design*. Doctoral thesis, Lund: Lund University.

Paula, D. D., Dobrigkeit, F., & Cormican, K. (2018). From team collaboration to product success-the domino effect of design thinking. *DS 91: Proceedings of NordDesign 2018*, 14th–17th August, Linköping, Sweden.

Pugh, S. (1991). *Total Design: Integrated Methods for Successful Product Engineering*. Wokingham: Addison-Wesley.

Rapitsenyane, Y. (2014). *Supporting SMEs Adoption of Sustainable Product Service Systems: A Holistic Design-led Framework for Creating Competitive Advantage*. A Doctoral thesis, Loughborough: Loughborough University.

Rapitsenyane, Y., & Bhamra, T. (2013). The place of sustainability through product service systems in manufacturing SMEs in Botswana: A Delphi study. *Proceedings of the 16th Conference of the European Roundtable on Sustainable Consumption and Production (ERSCP) & 7th Conference of the Environmental Management for Sustainable Universities (EMSU)*, 4–7 June 2013, ERSP Society, Instanbul, Turkey.

Roozenburg, N., van Breemen, E., & Mooy, S. (2008). A competency-directed curriculum for industrial design engineering. *DS 46: Proceedings of E&PDE, the 10th International Conference on Engineering and Product Design Education*, September, Design Society, Barcelona.

Ruele, V. T. (2015). *An Investigation into the Management of Change in Design and Technology: A Qualitative Inquiry Based on the Implementation of a New Curriculum for Senior Secondary Schools in Botswana*. A Doctoral thesis, Loughborough: Loughborough University.

Salari, M., & Bhuiyan, N. (2018). A new model of sustainable product development process for making trade-offs. *The International Journal of Advanced Manufacturing Technology*, 94(1), pp. 1–11.

Sarala, R. M., & Vaara, E. (2010). Cultural differences, convergence, and crossvergence as explanations of knowledge transfer in international acquisitions. *Journal of International Business Studies*, 41(8), pp. 1365–1390.

Sentsho, J., Maiketso, J. T., Sengwaketse, M., Ndzinge-Anderson, V. & Kayawe, T. (2009). *Performance and Competitiveness of Small and Medium Sized Manufacturing Enterprises in Botswana*. Gaborone: Bay Publishing.

Seperamaniam, T., Jalil, N. A. A., & Zulkefli, Z. A. (2017). Hydrostatic bearing design selection for automotive application using Pugh controlled convergence method. *Procedia Engineering*, 170, pp. 422–429.

Srivastava, M. (2018). New product strategy/innovation: Challenges and opportunities in emerging market. In A. Adhikari (Eds.), *Strategic Marketing Issues in Emerging Markets*. Singapore: Springer. https://doi.org/10.1007/978-981-10-6505-7_10.

Statistics Botswana. (2018). *International Merchandise Trade Statistics*. www.statsbots.org.bw/sites/default/files/IMTS%20JANUARY%202018Z.pdf.

Statistics Uganda Trade and Industry Sector. (2013). Statistical abstract. *Kampala: Uganda Bureau of Statistics*. https://www.mtic.go.ug/wp-content/uploads/2019/08/Trade-and-Industry-Sector-Statistical-Abstract-2017-2018.pdf.

Stock, G. N., Tsai, J. C. A., Jiang, J. J., & Klein, G. (2021). Coping with uncertainty: Knowledge sharing in new product development projects. *International Journal of Project Management*, 39(1), pp. 59–70.

Taheri, B., Pourfakhimi, S., Prayag, G., Gannon, M. J., & Finsterwalder, J. (2021). Towards co-created food well-being: Culinary consumption, braggart word-of-mouth and the role of participative co-design, service provider support and C2C interactions. *European Journal of Marketing*, 55(9), pp. 2464–2490. https://doi.org/10.1108/EJM-02-2020-0145.

Tan, A., McAloone, T. C., & Andreasen, M. M. (2006). What happens to integrated product development models with product/service-system approaches? *Proceedings of the 6th Workshop on Integrated Product Development*, September, Magdeburg. The Design Society.

Ulrich, K. T., & Eppinger, S. D. (2000). *Product Design and Development*. New York: McGraw-Hill Education.

Ulrich, K. T., & Eppinger, S. D. (2003). *Product Design and Development*. New York: McGraw-Hill Education.

Ulrich, K. T., & Eppinger, S. D. (2012). *Product Design and Development*. New York: McGraw-Hill Education.

Van Oers, B. (1998). From context to contextualising. *Learning and Instruction*, 8(6), pp. 473–488.

Vezzoli, C., Ceschin, F., Osanjo, L., M'Rithaa, M. K., Moalosi, R., Nakazibwe, V., & Diehl, J. C. (2018). Design for sustainability: An introduction. In *Designing Sustainable Energy for All*, pp. 103–124. Cham: Springer.

Von Hippel, E. (2005). Democratizing innovation: The evolving phenomenon of user innovation. *Journal für Betriebswirtschaft*, 55, pp. 63–78. https://doi.org/10.1007/s11301-004-0002-8.

Waage, S. A. (2007). Re-considering product design: A practical "road-map" for integration of sustainability issues. *Journal of Cleaner Production*, 15(7), pp. 638–649.

Zhu, X., Xiao, Z., Dong, M. C., & Gu, J. (2019). The fit between firms' open innovation and business model for new product development speed: A contingent perspective. *Technovation*, 86, 75–85.

4
HUMAN FACTORS AND UBUNTU IN PRODUCT DESIGN

Oanthata Jester Sealetsa and Patrick Dichabeng

Introduction

Human factors is a technical discipline concerned with understanding inter-linkages among humans and other system elements, and the occupation applies principles, theory, data and methods to design to optimise human well-being and overall system performance (Village & Neumann, 2014; International Ergonomics Association, 2000). However, human factors can be divided into three sub-disciplines: physical, cognitive and organisational. Physical human factors are concerned with the human body's limitations when performing a task, while human cognitive factors involve observing the limitations of the human mind when people perform tasks. On the other hand, human organisational factors are related to the setup of the organisation and how this setup can affect human performance. Therefore, the human factor is a scientific subject that attempts to harmonise humans with the working environment by matching people with tasks. This is chiefly to prevent the risk of accidents and injury. In the context of product design, knowledge and awareness of human factors can aid product designers in matching their designs with the capabilities of users to prevent accidents and injuries when products are used. However, human factors appear admirably established in well-developed counties, and little seems to be known about human factors in emerging economies. This may lead to limited use to support product design and occupational health and safety in these expanses. Several factors hindering human factors application and use have been reported in Afrika. For example, Olabode et al. (2017) observed an absence of knowledge of ergonomic design and its application in Nigeria. In South Africa, Scott and Charteris (2004), Christie (2012) and Thatcher et al. (2018) echoed similar

DOI: 10.4324/9781003270249-6

sentiments on the scarcity of human factor facilities. In Ghana, Pickson et al. (2017) identified many instances where human factors are sometimes nonexistent. In all these countries, human factors awareness, insufficient relevant human factors information, lack of trained personnel, lack of human factors researchers, adequate resources and a divide between employment and people are some of the various factors hindering the use and application of human factors in Afrika. It is noted that such circumstances could lead to low acceptance of human factors in product design and the proliferation of poorly designed products deficient in human factors in Afrika. There is a demand to increase awareness and acceptance of human factors in emerging economies so that poorly designed products that do not conform to health standards can be significantly reduced. To achieve this in Afrika, human factors practice must be based on the foundation provided by the philosophy of Ubuntu. To demonstrate how this arrangement can be fashioned, the authors draw inspiration from the traditional Afrikan mud-walled hut and propose a contextualised model shown as the Ubuntu model of human factors in Figure 4.2.

The traditional mud-walled hut is a round structure usually made of cheap construction materials such as mud or clay and thatch, with a peaked roof inspired by the Egyptian pyramids (Shoenleber, 2008). In essence, the hut has five different features, namely, the foundation, the pillars, the wall, the roof and the rooftop cone. These features have different functions. For example, the foundation keeps out moisture from the ground and supports the pillars holding the roof. The thatched roof keeps away water and makes the hut cool and warm depending on the outside temperature. However, the rooftop cone signifies the completion of the hut. It is usually where the thatch is bundled together for aesthetic purposes and to prevent rainwater from seeping into the house from the top. The result is a cheap but reasonably strong architecture whose significance can be found in its commonality and complementary nature and not in enslaving people with ownership, as with modern architecture (Shoen leber, 2008). This traditional mud-walled hut is shown in Figure 4.1.

The proposed model (Figure 4.2) illustrates the philosophy of Ubuntu as the foundation of the traditional mud-walled hut. This feature can represent the same function as the foundation of the traditional mud-walled hut depicted in Figure 4.1, acting as the ground support for the relevant human factors' principles. On the other hand, the three fields of ergonomics, namely, physical ergonomics (PE), cognitive ergonomics (CE) and organisational ergonomics (OE), illustrated as the pillars supporting the roof of the hut, can also represent the function of supporting the design processes shown in Figure 4.2 as the roof of the hut. However, the design processes are presented as the hut's thatched roof, while the rooftop cone is depicted as the products in Figure 4.2.

Therefore, Figure 4.2 illustrates how the philosophy of Ubuntu can be used as a support for ergonomics principles. These principles can then be

FIGURE 4.1 Traditional mud-walled hut

Source: Photograph by the authors

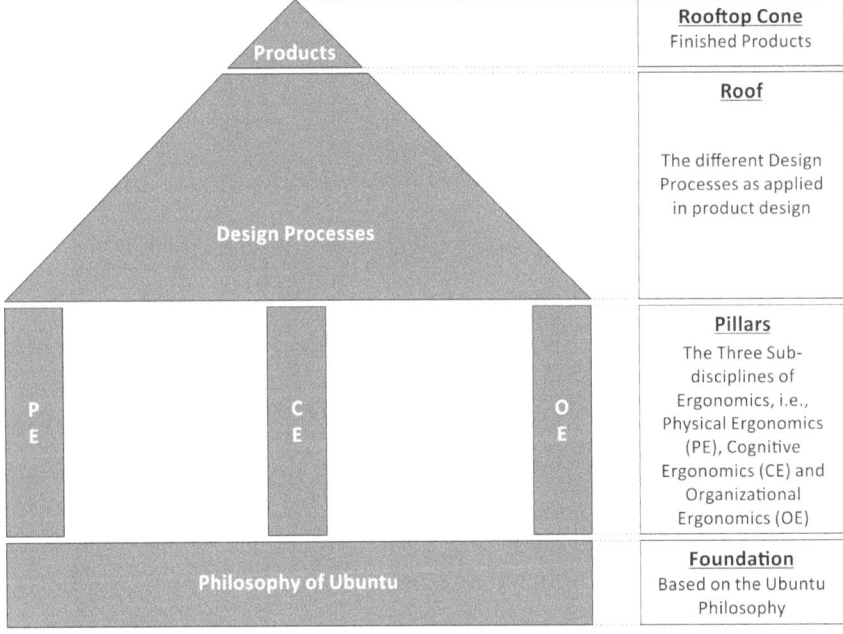

FIGURE 4.2 Ubuntu model of human factors

Source: Illustration by the authors

applied to inform product design processes. The arrangement can lead to the improvement of product design in industrially developing countries.

Strengthening Other Disciplines With Ubuntu

Subsequent to the Truth and Reconciliation Commission process's success in South Africa, evidence suggests that the Ubuntu philosophy has extensively been used to strengthen other areas of scholarly work. For example, in management, Mayaka and Truell (2021), in their study proposing a new agenda for social work and development, show how Ubuntu can support and strengthen international social work ethical principles and practice. In support of this argument, Lutz (2009) intensely observed that a theory of global management consistent with ordinary human nature comparable to some dominant Western ethical principles is needed because of globalisation. Lutz (2009) argues that some Western principles may not apply in Afrika's context. Moray (1995) called for a multidisciplinary approach based on the complexity of the world's interconnectedness, calling for a philosophy that could bind people's interconnectedness. Based on the previous observations, including that of Meiring (2015), the Truth and Reconciliation process in South Africa became the anchor of the philosophy. Ubuntu can also be extended to ergonomics because it has abundantly been shown that this philosophy can support many disciplines.

For example, in theology, Ubuntu has taught forgiveness, reconciliation, respect for human dignity and justice and recognising human beings' interconnectedness (Tutu, 2000). The theological Ubuntu argues that human beings have been created in resemblance of God and is based on Archbishop Desmond Tutu's Christian faith (Tutu, 2000; Meiring, 2015).

In business, Taylor (2014) perceives Ubuntu as a framework for ethical business decision-making, arguing that this philosophy can be applied to various disciplines. In corroborating this view, Hailey (2008) considers Ubuntu as a basis for business ethics, while Auchter (2017) sees Ubuntu as an Afrikan view of global business ethics. Furthermore, Benyera (2014) perceives Ubuntu as being utilised to inform chief executive officers on corporate social responsibility and codes of ethics in business. However, it appears in the discipline of business that Ubuntu's philosophy has gained extensive usage. In management, Mayaka and Truell (2021) show how Ubuntu can support and strengthen international social work ethical principles and practices.

In computer science, the term Ubuntu has become synonymous with Linux computers. These are computer operating systems used to distribute software. Linux operating systems are open source, meaning everyone can freely use them, and they are based on the concept of sharing, which is quite rare in the closed world of computer software distribution. The software was designed by Mark Shuttleworth, a South African–born British entrepreneur who, in 2002, became the first South African to travel as a tourist to space (Tabassum & Mathew, 2014). The Linux computer systems have been inspired by the philosophy of Ubuntu, whose name has been derived, creating global interest in the concepts and values implicit in the term Ubuntu (Tabassum & Mathew, 2014; Anusha & Devaki, 2020).

In public health, Ubuntu has recently been used in the fight against COVID-19. For example, Chingangaidze et al. (2021) found that Ubuntu can provide ways to deal with challenges that emerge with the COVID-19 pandemic. Sambala et al. (2019) argue that Ubuntu could be a valuable philosophy for negotiating ethical decisions arising from disagreements from moral tensions between demands for civil liberties and goals of public health. In citing the challenges that may be created by vaccine hesitancy, one of which is delaying countries reach herd immunity thresholds of coverage against COVID-19, Ndwandwe and Wiysonge (2021) cite vaccine hesitance as the most significant challenge that can be minimised by incorporating Ubuntu in education about the need to vaccinate. Other areas that have used Ubuntu to advance their course are nursing (Nolte & Downing, 2019), psychology (Hanks, 2007; Wilson & Williams, 2013) and medicine (Washington, 2020).

The mentioned examples illustrate how other disciplines have proceeded to exploit and profit from the philosophy of Ubuntu. Therefore, its appendage as an anchor for human factors in product design in Afrika can be a welcome development. This is because human factors have some of its tenets that overlap with the philosophy of Ubuntu. This overlap has to be scrutinised critically and discussed. Therefore, what follows are discussions on the tenets of the philosophy of Ubuntu and the foundation they create on which human factors can leverage.

The Influence of Ubuntu on the Application of Human Factors

Ethical considerations are the first tenet of the philosophy of Ubuntu that can influence the application of human factors. As has already been shown in business, ethically conducting business has become an integral part of modern business. Equally so, there is a need to conduct human factors research ethically. For example, it is essential to consider society's cultural practices in researching human factors. In Botswana, for example, it could be challenging to collect anthropometric data because of cultural beliefs and practices driven by myths and witchcraft. Convincing potential users to provide anthropometric data of the feet, head and waist might challenge the researchers, as taking measurements of these body parts is associated with witchcraft and certain myths. For example, '*go tsaa lonao*' can be misconstrued as gathering information about somebody's feet for witchcraft. Understanding and respecting these beliefs and involving village chiefs and elders using the Kgotla system, where people can talk and consult freely about something they do not like, can assist human factors researchers around the challenges. Collecting and compiling anthropometric data following established protocols is an important ethical consideration. Most emerging economies do not have anthropometric data on their citizens.

In most cases, Western anthropometric data is used, which mismatches local users because of differences in body sizes. As alluded to before, collecting such data is related to myths, and it can be a daunting task to achieve. Therefore, applying the philosophy of Ubuntu can overcome the hurdle.

According to Metz (2011) and Tutu et al. (2000), Ubuntu promotes human rights and the dignity of others. On the other hand, ergonomics deals with health and safety issues, such as the prevention and management of musculoskeletal disorders (Bridgers, 2008). These ailments may take away human dignity by reducing people's ability to function freely without being aided. Human factors also promote the right for workers to know hazards, risks and protective laws to prevent injuries in the workplace that can, in the future, influence the degradation of people's dignity by sustaining injuries that may have resulted from poor human factors. Ubuntu also focuses on respect for the dignity of others and may also offer people a good working environment. Similarly, human factors promote good and healthy working environments (Sealetsa & Thatcher, 2011; Olabode et al., 2017; Öztürk & Arici, 2018).

Another doctrine of Ubuntu relevant to the study is the doctrine of teamwork, which promotes consensus and dialogue among team members. In human factors, teamwork is essential because work systems are based on the performance of teams (Paris et al., 2000; Sexton et al., 2000; Xiao et al., 2013). A philosophy that encourages teamwork can enrich human factors, particularly in aviation and health care, where teamwork performance is paramount as it helps workers to cooperate and protect each other (Xiao et al., 2013). Furthermore, communication channels can easily be created and supported because of teams' trust and interdependence. In this regard, accidents and injuries can be reduced in the workplace.

Ubuntu can further promote harmony between people, the environment and technology. In this regard, Ubuntu can encourage ergonomists and product designers to understand further the need for a fit between people, equipment and the environment. In Afrika, ergonomics is almost nonexistent and is often driven by the financial opportunities that may not be available in developing countries (Thatcher et al., 2018).

Another important tenet of Ubuntu that resonates with human factors is social responsibility. In this principle, Ubuntu relates to understanding people's needs as they relate to the environment. Product designers are also responsible for keeping the environment safe from climate change, deforestation and pollution. Therefore, responsibility for human factors lies in making environmental considerations when designing artefacts. In this regard, human factors can be applied from a humanistic perspective where designers can be morally obliged to be socially responsible by promoting universal design, user-centred design and processes, environmental friendliness and eco-friendly designs (M'Rithaa, 2011).

Limitations of Ubuntu as an Influence on Human Factors Application

While Ubuntu can offer support for human factors advancement, several researchers, such as Metz (2011), van Hensbroek (1999), Louw (2006) and Broodryk (2005), offer caution on the possible pitfalls of Ubuntu. For example, van Hensbroek (1999) describes Ubuntu as a philosophy that dwells on what he refers to as 'bipolar thinking', where Afrikans are pitted against the West because of their different beliefs and cultural aspirations. Van Hensbroek (1999) even doubts the existence of Ubuntu and further posits that Ubuntu is a practice that blinds the reality of facts to people. On the other hand, Louw (2006) views Ubuntu as ancient wisdom in modern work. Metz (2011) explains the vulnerability of Ubuntu to objections, such as Ubuntu not fitting modern, industrial society. Broodryk (2005) believes that Ubuntu is a doctrine not premised on empirical evidence and is not even recorded because it is anchored on Afrikan rituals, customs and practices. Broodryk (2005) contends further that such a scenario creates an opportunity for foreign ideologies to challenge Ubuntu. Therefore, anchoring a highly scientific discipline, such as human factors, on a doctrine that is not adequately documented might refute and compromise the ethical considerations required in research. However, as Afrikan society can be humanistic, perhaps using Ubuntu to improve product acceptance in design. Furthermore, it can foster responsibility and the seriousness of understanding responsibility in design.

Conclusion

Based on the understanding and appreciation that Ubuntu has been used to influence other disciplines such as business, management, and medicine, the authors believe that there is a role in human factors that Ubuntu can play. However, more research on the philosophy of Ubuntu as a possible anchor for human factors is needed. In Botswana, ergonomics is very expensive and sometimes not practised at all. At times the laws governing its application, such as the Factory's Act in Botswana, are fragile and, therefore, subjected to abuse (Sealetsa & Thatcher, 2011). Given this scenario, Ubuntu can fill that void and make product designers appreciate the value of human factors in product design. Moray (1995) contends that the world needs a customary guiding principle of human values, and Ubuntu might be that common principle.

Reference List

Anusha, S., & Devaki, P. (2020). Development of Ubuntu performance monitoring tool. *International Journal of Science Research in Computer Science, Engineering, and Information Technology*, 6(3). ISSN 24563307.

Auchter, L. (2017). An African view on global business ethics: Ubuntu – A social contract interpretation. *International Journal of Business and Economic Development (IJBED)*, 5(2).

Benyera, E. (2014). Exploring Zimbabwe's traditional, transitional justice mechanisms. *Journal of Social Sciences*, 41(3), pp. 335–344.

Bridgers, R. (2008). *Introduction to Ergonomics*. Florida: CRC Press.

Broodryk, J. (2005). *Enterprise Risk Assessment-A New Approach for a Tough Environment*. https://phys.org/news/2022-01-approach-enterprise.html [Accessed 11 January 2006].

Chigangaidze, R. K., Matanga, A. A., & Katsuro, T. R. (2021). Ubuntu philosophy as a humanistic – Existential framework for the fight against the COVID-19 pandemic. *Journal of Humanistic Psychology*, 00221678211044554.

Christie, C. J. A. (2012). Straightforward yet effective ergonomics collaborations in South Africa. *Ergonomics in Design*, 20(4), pp. 39–42.

Hailey, J. (2008). *Ubuntu: A Literature Review*. London: Tutu Foundation.

Hanks, T. L. (2007). The Ubuntu paradigm: Psychology's next force? *Journal of Humanistic Psychology*, 48(1), pp. 116–135.

International Ergonomics Association Congress, Human Factors, & Ergonomics Society Meeting. (2000). *Ergonomics for the New Millennium: Proceedings of the XIVth Triennial Congress of the International Ergonomics Association and the 44th Annual Meeting of the Human Factors and Ergonomics Society*, July 29 Through August 4, 2000, San Diego, CA (Vol. 1). Human Factors and Ergonomics Society.

Louw, D. J. (2006). The African concept of ubuntu and restorative justice. In D. Sullivan & L. Tifft (Eds.), *Handbook of restorative justice: A global perspective*, pp. 161–174. New York: Routledge.

Lutz, D. W. (2009). African Ubuntu philosophy and global management. *Journal of Business Ethics*, 84(3), pp. 313–328.

Mayaka, B., & Truell, R. (2021). Ubuntu and its potential impact on the international social work profession. *International Social Work*, 64(5), pp. 649–662.

Meiring, J. J. (2015). Ubuntu and the body: A perspective from theological anthropology as embodied sensing. *Verbumet Ecclesia*, 36(2), pp. 1–8.

Metz, T. (2011). Ubuntu as a moral theory and human rights in South Africa. *African Human Rights Law Journal*, 11(2), pp. 532–559.

Moray, N. (1995). Ergonomics and the global problems of the twenty-first century. *Ergonomics*, 38(8), pp. 1691–1707.

M'Rithaa, M. K. (2011). *Universal Design in Majority World Contexts: Sport Megaevents as Catalysts for Social Change*. Sunnyvale: Lambert Academic Publishing.

Ndwandwe, D., & Wiysonge, C. S. (2021). COVID-19 vaccines. *Current Opinion in Immunology*, 71, 111–116.

Nolte, A., & Downing, C. (2019). Ubuntu – the essence of caring and being: A concept analysis. *Holistic Nursing Practice*, 33(1), pp. 9–16.

Olabode, S. O., & Adesanya, A. R. (2017). Ergonomics awareness and employee performance: An exploratory study. *Economic and Environmental Studies*, 17(4(44)), pp. 813–829.

Öztürk, D., & Arici, Y. K. (2018). A study on workplace environment ergonomics and conditions of employees in rice factories. *Business Studies*, 41.

Paris, C. R., Salas, E., & Cannon-Bowers, J. A. (2000). teamwork in multi-person systems: A review and analysis. *Ergonomics*, 43(8), pp. 1052–1075.

Pickson, R. B., Bannerman, S., & Ahwireng, P. O. (2017). Investigating the effect of ergonomics on employee productivity: A case study of the butchering and trimming line of pioneer food cannery in Ghana. *Modern Economy*, 8(12), p. 1561.

Taylor, D. F. P. (2014). Defining ubuntu for business ethics – a deontological approach. *South African Journal of Philosophy*, 33(3), pp. 331–345.

Tutu, D. (2000). *No Future Without Forgiveness: A Personal Overview of South Africa's Truth and Reconciliation Commission*. London: Rider Random House.

Sambala, E. Z., Cooper, S., & Manderson, L. (2019). Ubuntu as a framework of ethical decision-making in Africa. *Ethics & Behavior*, 30(1), pp. 1–13. https://doi.org/10.1080/10508422.2019.1583565.

Scott, P. A., & Charteris, J. (2004). Ergonomics in industrially developing countries (IDCs): Socio-cultural perspectives. In *Cultural Ergonomics*. Bingley: Emerald Group Publishing Limited.

Sealetsa, O. J., & Thatcher, A. (2011). Ergonomics issues among sewing machine operators in the textile manufacturing industry in Botswana. *Work*, 38(3), pp. 279–289.

Sexton, J. B., Thomas, E. J., & Helmreich, R. L. (2000). Error, stress, and teamwork in medicine and aviation: Cross-sectional surveys. *BMJ*, 320(7237), pp. 745–749.

Shoenleber, L. (2008). *Why Africans Live in Huts*? https://ezinearicles.com [Accessed 5 July 2022].

Tabassum, M., & Mathew, K. (2014). Software evolution analysis of linux (Ubuntu) OS. *International Conference on Computational Science and Technology*, ICCST, pp. 1–7, Kota Kinabalu, Malaysia. https://doi.org/10.1109/ICCST.2014.7045194.

Thatcher, A., Waterson, P., Todd, A., & Moray, N. (2018). State of science: Ergonomics and global issues. *Ergonomics*, 61(2), pp. 197–213.

Tutu, A. E. D., Tutu, D., Botman, H. R., Botha, M. E., Degenaar, J., De Grunchy, J. W., & Smith, R. D. (2000). *Race and Reconciliation in South Africa: A Multicultural Dialogue in Comparative Perspective*. Pennsylvania: Lexington Books.

van Hensbroek, P. B. (1999). *African Renaissance and Ubuntu Philosophy*. Groningen: University of Groningen.

Village, J., & Neumann, W. (2014). Designing for human factors. *Industrial and Systems Engineering at Work*, 46(11), p. 36.

Washington, K. (2020). Journey to authenticity: African psychology as an act of social justice honouring African humanity. *Journal of Humanistic Psychology*, 60(4), pp. 503–513. https://doi.org/10.1177/0022167820917232.

Wilson, D., & Williams, V. (2013). Ubuntu: Development and framework of a specific model of positive mental health. *Psychology Journal*, 10(2).

Xiao, Y., Parker, S. H., & Manser, T. (2013). Teamwork and collaboration. *Reviews of Human Factors and Ergonomics*, 8(1), pp. 55–102.

PART 2
Design Research

5

DESIGN-DRIVEN RESEARCH METHODS

Richie Moalosi and Caiphas Thusonyana Othomile

Introduction

Many methods, approaches, and tools can be used at different stages of the design process. If designers know such, they can select the most suitable methods for each step to achieve the desired outcomes. Milton and Rodgers (2013) argue that research-driven methods are somewhat neglected globally in product design. This chapter intends to fill this gap by introducing design students and practitioners to various practical research methods and tools throughout the design process. Some examples of how these tools are used, with samples of design research applying the same, have been provided.

Design-driven research aims to generate utility value for the stakeholders by systematically gathering human experiences that are synthetically processed to develop a comprehensive solution that meets the identified need (Norman & Verganti, 2014). The methods, tools, and approaches used in design-driven research provide insights that inform problem-solving. Insights may include revelations on why people currently do what they do and assumptions about how their experiences can be improved. The data obtained from design-driven research help designers extrapolate initial design concepts to work on, that is, how the problem can be solved based on the information gathered and processed. Design-driven research challenges provoke and disrupt the status quo and produce trans-disciplinary and heterogeneous knowledge that strives to improve the welfare of stakeholders (Milton & Rodgers, 2013). It provides a basis for evidence-based decision-making to develop products or services that meet stakeholders' needs they are yet to articulate to themselves. The chapter discusses various design tools used in product or service development, such as inquiry, define, design selection, prototyping, and testing.

DOI: 10.4324/9781003270249-8

For example, the Ubuntu co-creation process is an authentic, decolonial, and indigenous approach to design. This chapter will replace the commonly used word 'user' with 'stakeholders.' The word 'user' is associated with traditional linear manufacturing processes, which is unsustainable because it is limited to the end user. Since the world is transitioning to a circular economy, a much broader term such as 'stakeholders' considers people who extract resources, build, use, and dispose of products, services, and systems.

Design-Driven Research

Design-driven research involves using various research methods that investigate human experiences and behaviour. The outcomes inspire designers to develop new insights and ways to address stakeholders' needs by designing pleasurable products, services, and systems. It is specifically undertaken to improve the strategic design and development of products, services, and systems. It allows designers to understand complex human behaviour and turn that into actionable insights to improve the design. The goal is to enable designers to design with their stakeholders. This approach provides a vehicle for observation, reflection, interpretation, discussion, and expression about future ways of living (Nelson & Stolterman, 2012; Faste & Faste, 2012; Sanders & Stappers, 2014). Design-driven research is a human-focused method that helps designers to answer questions such as:

- Who are the stakeholders in the project undertaken?
- What challenges are they facing?
- How will they use the proposed product, service, or system?

Qualitative and quantitative research aims to find design inspiration based on how stakeholders use the proposed product or service. There are two research techniques in design-driven research.

1 Primary research: This method involves the designer's original study to collect new data by directly engaging with the stakeholders by observing, learning, and asking questions. A range of techniques, such as observations, empathy tools, focus groups, usability sessions, surveys, and interviews, are used to collect data.
2 Secondary research: This involves using existing research from books, journal articles, sales reports, government statistics, renowned magazines, or online sources to authenticate existing research. Designers use it to create a stronger case for their design choices and provide additional insights into what they have learned during primary research. Occasionally, it can be used to shape and drive primary research.

Benefits of Design-Driven Research

Though design-driven research takes time, resources, and preparation, the results lead to the satisfaction of stakeholders' needs by designing desirable products or services. Some of the benefits of design-driven research include the following:

1 Allowing designers to design products or services based on facts and not assumptions. In some instances, designers may think that they know their stakeholders. However, the designer can only understand what stakeholders need and value; their abilities, pain points, and limitations by interacting with them; their expectations; and how they will use the product or service. Assumptions lead to designing the wrong products, services, or systems. Therefore, designers need to observe people in their natural environment, talk to them, interact with them, feel what they feel, learn from them, and gain empathy for stakeholders.
2 Helping designers to focus on the task and prioritise tasks to be undertaken. The data from stakeholders will help the designer focus on and prioritise important tasks.
3 Nurturing stakeholders' empathy. Interacting with stakeholders helps designers capture stakeholders' thoughts and feelings, thus building a deep understanding of stakeholders' needs.
4 Leading to customised solutions.
5 Creating design impact, thus designing desirable, feasible, and viable products, services, and systems. In support of the latter, Daalhuizen (2014) argues that methods are a means to assist designers in attaining desired change as efficiently and effectively as possible.
6 Improving stakeholder satisfaction.
7 Contextual and holistic understandings of stakeholder experiences can inform value propositions that better fit stakeholders' value-in-use (Yu & Sangiorgi, 2017).

However, Norman and Verganti (2014) argue that design-driven research does not lead to radical but incremental innovation. Radical innovation is driven by technology or meaning changes without design-driven research or formal analysis of a person's or even society's needs. Examples of radical innovation include Google, Facebook, and Twitter.

Ubuntu research-driven research is based on authentic Afrikan values of viewing oneself through others. It promotes interconnectedness between people (designers and the community) – thus, one's humanness and personhood depend upon one's relationship with others. Before research can commence, designers must build genuine relationships with the community rather than using them as research objects. Such an approach provides Afrikan designers with ways to address challenges they experienced using Euro-centric design tools. The Ubuntu design approach integrates human components such as psychological, biological, spiritual, and environmental facets of life when addressing society's challenges (Chigangaidze et al., 2022). The tools promote a coordinated approach to societal responsibility, generosity, teamwork, and sharing. Tools may be the same as the Euro-centric ones, but they should be used with a decolonised point of view that considers the Afrikan people and context.

Design-Driven Methods

This chapter focuses on the different methods designers can use in primary research. It involves the designers or design teams going directly to the source (stakeholders) to observe, ask questions, and gather data. The goal is to understand better who the stakeholders are that they are designing for or to validate their ideas with the actual stakeholders. The design-driven methods for different phases of the design process have been divided as follows: inquire, define, design, prototyping, testing and evaluation, and communication (Table 5.1).

Inquire Methods

The inquiry phase aims to analyse the problem space and help the designer or design team understand the wicked problems stakeholders are experiencing. Such an analysis gives designers a more profound personal insight into the problem and points out how the same can be resolved. As shown in Table 5.1, many tools can be used to understand the situation better. Afrika is part of the global village, and some design tools are borrowed from the Global North and then contextualised to the continent's needs. This section discusses tools which emphasise empathy as one of the tenets of Ubuntu philosophy, such as co-creation, in-context immersion, empathy stakeholder observation, and empathy interviews. Empathy enables designers to see stakeholders' problems through their eyes, feel what they feel, and have a first-hand experience of the wicked problem under investigation. The process involves observing, engaging, and empathising with the stakeholders the designer is designing for to understand their experiences, motivations, needs, and challenges.

Ubuntu Co-Creation

Co-creation (collective creativity) is when stakeholders creatively collaborate with designers or other stakeholders to produce innovative systems, products, and services of value in the marketplace (Moalosi et al., 2019). Such an approach to design has provided a new paradigm shift in how stakeholders are invited to design their systems, products, and services. Therefore, stakeholders

TABLE 5.1 Design-driven research methods

Inquire	Define	Design	Selection Evaluation	Prototyping	Testing and Evaluation	Communication
Empathy stakeholder observation, Cultural probes, Context mapping, Empathy interviews, Questionnaires, Stakeholder's journey map, Mind mapping, Focus groups, Co-creation, In-context immersion, Product autopsy, etc.	Personas, Mood board, Product collage, Storyboard, Business model canvas, Written scenario, Problem definition, SWOT analysis, Empathy maps, etc.	Analogies and metaphors, Biomimicry, Synectics, SCAMPER, SCAMMPERR, Bisociative technique, Brainstorming, Brainsketching, Who-what-where-when-why-how, Six thinking hats, Lateral thinking, TRIZ, 2D and 3D CAD, etc.	Datum method, C-Box, VALUE, Product concept evaluation, Product usability evaluation, SWOT, New, Appeal and Feasibility, Controlled convergence, etc.	Sketch modelling, Rapid prototyping, 3D CAD simulations, mock-ups, Paper prototyping, Appearance models, Quick-and-dirty prototypes, Experience prototyping, etc.	Iterative testing, Expert testing, Laboratory stakeholder testing, Guerilla testing, Co-discover, Think-aloud protocol, etc.	Report writing, Presentation, Product posters, Demonstration videos, Stakeholder manuals, etc.

want to create meaning, value, and relationships by designing products and services. This contrasts with the conventional practices that rendered stakeholders largely passive in value creation. They were researched, observed, segmented, targeted, marketed, and in the end, products were sold to them. The principles of co-creation include producing value for stakeholders to participate passionately. In the co-creation process, value is co-produced by focusing on stakeholder experiences. Stakeholders are provided with a forum to interact and share their experiences through the process directly.

Design teams use co-creation because it enables them to formulate breakthrough design and development strategies (better innovation), design compelling new products and services, and transform management processes. Moreover, the method allows designers to lower risks and costs, increase market share, thus improving competitiveness, loyalty, and returns, and generating more value than traditional transactions. There is also high perceived quality in co-created products and higher stakeholder satisfaction. Stakeholders should be involved throughout the process, from discovery to testing and evaluation. The design team or designer should co-create or design with the stakeholders. Stakeholders are equal partners as they are experts in their experiences. Stakeholders' involvement in the co-creation process builds trust in the developed product or service.

From an Afrikan perspective, the concept of co-creation in design has a similar ethos to the philosophy of Ubuntu, thus Ubuntu co-creation. Ubuntu, translated from several Bantu languages, means humanness or "I am because we are and since we are, therefore, I am" (Olasunkanmi, 2015). Ubuntu defines a process for gaining respect by first giving it and gaining empowerment by empowering others. Everyone is invited to contribute towards the goals and commonwealth of the community. Inclusive decision-making and participatory meetings are critical traditions in Afrikan communities. If the community is empowered, it can be a source of competence that can co-create creative and contextually appropriate innovations. Suppose the co-creation process is merged with the philosophy of Ubuntu. In that case, the method may lead to co-creating innovative, effective, and sustainable solutions that are contextually relevant, culturally appropriate, and inclusive designs rooted in the Afrikan or stakeholders' values.

One of the critical stages in the Ubuntu co-creation, which is not found in other design processes, involves preliminary visits to gain access to the stakeholders, develop rapport, and build trust and empathy with the community or stakeholders (Moalosi et al., 2017; Chilisa et al., 2016). The process involves getting a gatekeeper to introduce the design team to the community and meeting the village elders or chief to explain the project objectives and outcomes. This process helps the design team understand the community's real needs through direct interaction. This stage is often ambiguous and chaotic and involves asking open-ended questions describing the problem statement. The community identifies the problem collectively by sharing their indigenous knowledge, life experiences, and social needs. The design team must understand the problem context, community customs, norms, beliefs, values, and spirituality (Moalosi et al., 2017). The community's social needs become the design team's needs. During the discovery phase, empathy is cultivated to hear and feel what others share from a deeper perspective. Ubuntu's philosophy encourages sharing generously with others. The 'I' is eliminated, and the 'we' state of mind becomes predominant in proposing design ideas. A shared vision, caring, consensus decision-making, trust, respect, and dignity are some of the tenets of the philosophy of Ubuntu, which should be displayed during the co-creation process (Moalosi et al., 2017).

When the project is complete, the design team must develop an exit strategy to ensure sustenance. Often, communities are fatigued with being used as research objects on projects which do not benefit them. An exit strategy should be gradual to ensure the community assumes ownership and responsibility for the project. The community should be empowered to sustain the project on its own. All in all, embracing the spirit of Ubuntu in the co-creation process enhances more meaningful stakeholder participation and the creation of valuable products, services, and systems. Taking on board the spirit of Ubuntu into design thinking recognises non-Western ways of thinking rooted in craft practices. It enhances more meaningful participation where socio-economic access to technological innovation and the epistemological circumstances of designers and stakeholders differs acutely (Winschiers-Theophilus et al., 2012).

In-Context Immersion

The best way for designers to understand a group, subculture, setting, or method of life is to immerse themselves in that context or become part of the group or topic of study. Meeting stakeholders where they live, work, and socialise provides designers with new insights and unexpected opportunities (IDEO, 2015). By being with people in their natural settings and doing what they usually do (culture), the designer can talk about their experiences. It also means listening to the people and attempting to see the world from their point of view. This enables designers to gain empathy and understand the stakeholders they are designing with on an intellectual and experiential level, as Brand and Campbell (2014) demonstrated when designing a household farming kit for small-scale farming.

It is an excellent way to get to know the landscape of the design challenge and experience it from the stakeholder's point of view. Stakeholders' culture consists not just of the physical environment but also ideologies, values, and ways of thinking. Designers can get an in-depth and longitudinal understanding of the subject. This understanding helps designers gain deeper insights, identify hidden needs, and design solutions from stakeholders' perspectives (Liedtka, 2018). Such knowledge inspires the design, changes behaviour, and builds trust. Stakeholders are more than willing to share their experiences once they feel that the designer cares about what they are offering, shows empathy, and shares their knowledge, experiences, etc., to create a balanced relationship. Suppose the designer pays attention to this trust-building period. In that case, it will be easier in the interview process to get honest answers, and the designer may even make some stakeholders uncomfortable.

The designer should know the stakeholders' cultural norms when conducting in-context immersion. If this is neglected, it could affect the interview or observation process. The designer should respect and be down to earth when dealing with stakeholders. The designer also needs to be open to new ideas and influences because there is always an opportunity to learn from the stakeholders and their context. Immersing oneself in the stakeholders' world is open to new creative possibilities and allows the designer to leave behind preconceived ideas and outdated ways of thinking (Kolawole, 2015). The designer needs to listen, observe, analyse, and design sustainable solutions.

Depending on the project, observing stakeholders for a few hours to a week is ideal. It is essential to record or take down details of what one sees and hears. Taking quotations on what stakeholders say strengthens the designer's story by providing first-hand concrete evidence. Alternatively, one can shadow a stakeholder for some time and, in the process, ask the stakeholder about their life and how they make decisions and watch them socialise, work, and relax (Brand & Campbell, 2014; IDEO, 2015). In summary, in-context immersion opens opportunities to understand what goes on under the surface and subconscious level and what the stakeholders being observed and interviewed may be experiencing. It is a flexible, unpredictable gathering that needs to drift naturally in the participants' direction. It is an ideal method to identify unknown pain points for stakeholders. However, one of the disadvantages of in-context immersion is that it requires a difficult balance between interviewing and observation. It is time-consuming and takes a great deal of dedication and finances. In most instances, when the researcher runs out of time, participants may be rushed through or skip some parts of the process to try to fit everything within the given time. In some work situations, in-context immersion can get out of control and become a session where participants complain about a product or application. A broader stakeholder should validate any insights from in-context immersion.

Empathy Observations

Observations are among the oldest and most fundamental research methods used in Afrika design. Innovators will observe a challenge and then develop a solution to address the same. This was done at an empathy level to uplift the community. Empathy requires designers to put aside their learning, culture, knowledge, opinions, and world view to understand other people's experiences of things deeply and meaningfully (Dam & Siang, 2020). Such an approach assists designers in gaining a deep understanding of the stakeholders who will use the design. Therefore, empathy stakeholder observations watch stakeholders in real (natural) settings as they interact with products, services, or systems and write down every behaviour they engage in. It allows the designer to understand what behaviour occurs when using a product or service. Dam and Siang (2020) argue that there are qualities and characteristics that designers need to develop to be more empathic observers, such as:

1 Abandoning one's ego – to empathise deeply, designers need to tame and put aside their egos and understand and experience the feelings of others.
2 Adopting humility – through humility, designers can elevate the value of others above themselves. They should be willing to admit their shortcomings and discard preconceived ideas for the benefit of the overall vision and goals.
3 Being a good listener – designers must listen attentively to uncover more profound meaning and stakeholders' experiences.
4 Refining observation skills – designers must develop reading stakeholders' behaviours, subtle indications, non-verbal expressions, body language, facial expressions, and voice intonations. The positive and negative signs from these expressions can lead to a deeper understanding of stakeholders' experiences.
5 Caring – designers must create a sense of care, a deep concern, and a desire to help, nurture, and assist their stakeholders.
6 Curiosity – by being curious, designers are prone to uncover unexpected areas, uncover new insights, and explore aspects of stakeholders' lives that help problem solving.
7 Genuineness – lack of sincerity kills empathy because it can be a barrier between designers and stakeholders. Designers should avoid approaching stakeholders with a superficial agenda, superiority complex, or mindset that may undermine their sincere intention to understand stakeholders' experiences deeply.

Developing the aforementioned traits will assist designers in discovering design opportunities they may never have imagined if they had only administered questionnaires. Observing stakeholders doing real work in real time enables designers to paint an accurate picture of how stakeholders interact with a product, service, or system (Martin et al., 2012). This enables designers to understand how a product, service, or system performs under natural conditions. Observations are usually a flexible data collection method, allowing the designer to collect information without burdening the stakeholders. A sample observation protocol for a study involving designers and leather manufacturing SMEs can be appreciated in Table 5.2 (adapted from Rapitsenyane, 2014).

TABLE 5.2 Sample observation protocol (adapted from Rapitsenyane, 2014)

[Observation schedule for all activities is mainly based on Barton et al. (1980) as cited in Robson (2011)]

Activity aims:
- To stimulate thinking differently
- To cultivate a positive perception of waste material
- To encourage group members to value others' thinking

Observations:
- The ability to work independently and when to engage other people
- The orientation of the outcome away from what will be expected
- Difficulties
- How different items are being utilised
- Are the results different?

Independent behaviours (how are they getting on, on their own?)
Dependent behaviours (where did group members request assistance? How was it offered and accepted?)
Independence-supportive behaviours (how group members encourage each other and how they discourage no-attempts from engaging in tasks)
Dependence-supportive behaviours (are designers doing everything and helping SMEs with everything without giving them a chance to go? Do designers discourage SMEs from equally engaging in tasks with little or no guidance from them?)

Their discussions or actions are not interrupted but recorded using video, pictures, and manual observation.

However, observations measure only what the designer can see. This method cannot collect other types of data (e.g., opinions, reasons behind behaviour). It is time-consuming, as one may need to conduct multiple observations to understand the stakeholders better. The results from observation cannot be generalised to other contexts. There is an element of the designer's bias when reporting the findings. Despite these limitations, observations are valuable for designers to gain stakeholder empathy.

Two types of observations are direct (reactive) and unobtrusive. In direct observation, stakeholders are aware that the designer is watching them. It involves observing them and recording them (either manually, electronically, or both). The only risk is that they are reacting to you. Individuals will change their actions rather than show the designer what they are like. However, contrived behaviour is challenging to maintain over time. Long-term observational studies will often catch a glimpse of natural behaviour. Ethically, if people see the designer and know they are watching them, they can ask them to stop the observation session.

Unobtrusive observation involves studying behaviour where stakeholders do not know they are being observed (Sauro, 2015). There is no concern that the designer may change the stakeholder's behaviour. Ethical issues on informed consent and invasion of privacy are paramount in observation sessions. In this method, the designer may pretend to join or become a member of a group and record information about the group. The group is unaware that they are observed for research purposes. The designer may take on several roles: complete participant, observer, participant, observer, and complete observer (Sauro, 2015).

Ross (2018) suggests that empathy observations can be conducted using the following levels of 'what, who, where, how, when, and why' (Table 5.3). However, the designer should try to observe through all senses, not just sight.

Another observational framework used is people, objects, environments, messages, and services (POEMS). POEMS is a simple framework that assists designers in conducting insightful and profound observational sessions (Table 5.4).

Activity, environment, interaction, object, and user (AEIOU) can also be used. AEIOU has five elements (Table 5.5).

Regardless of which observational method is used, one must analyse the data, identify themes and patterns emerging from the data, and produce an observation report on the findings that will pinpoint the pain points of stakeholders' experiences.

TABLE 5.3 Conducting empathy observations

Level	Description
What	a Under this level, the designer must write down details of what is happening, e.g., what is the stakeholder doing? What is happening in the stakeholder's context?
	b What are the emotions experienced by the stakeholders observed?
Who	a Find out who performs those tasks under observations.
Where	a Describe the context or location of the problem.
How	a Describe how the stakeholder is doing what they are doing – for example, observing a furniture designer using much energy to drive a screw into a wooden base?
	b Read the non-verbal expressions, body language, facial expressions, etc. Is the stakeholder using the right tools correctly for the job?
	c Describe the stakeholder's emotions and behaviours in performing the task.
When	a When are stakeholders performing those tasks or experiencing the problem? Is the problem continuous, or does it happen occasionally?
Why	a The designer needs to understand the stakeholder's context deeply. Why do stakeholders behave in a particular manner when performing specific activities?

TABLE 5.4 POEMS

Level	Description
People	Write the demographics, roles, behavioural traits, and the number of people observed.
Objects	What are the items people interact with within their environment? This could include furniture, devices, machines, appliances, tools, etc.
Environments	Describe the stakeholders' observations, e.g., architecture, lighting, furniture, temperature, atmosphere, etc.
Messages	Describe the social/professional interactions and environmental messages. What is the tone of the language or commonly used phrases in tag lines by participants?
Services	Outline all participants' services, apps, tools, and frameworks. Include any general comments observed during the session and supplement direct observations with photos or videos where appropriate.

TABLE 5.5 AEIOU

Level	Description
Activities	These are goal-directed sets of actions; people want to achieve. What ways do people use to work in specific activities and processes they do?
Environments	Describe the context of the activities, especially the atmosphere and function of the overall space, personal space, and shared space.
Interactions	Provide the building blocks of activities between people and their environment. Describe the nature of unique interactions between people and between people and objects in their context.
Objects	Provide the building blocks of the environment, key components that subject things to unintended uses, thus, changing their function, meaning, and context. Describe how people's objects and devices are related to the environment and their activities.
Users	State who the stakeholders are. These are the people whose behaviours, preferences, and needs are observed. Explain their roles and relationships as well as their values and prejudices.

Empathy Interviews

Any method promoting verbal conversation will yield better results in Afrika because of the continent's deep roots in oral traditions passed from one generation to the next. Ubuntu exemplifies the interconnectedness of all things and beings and interpersonal relationships (Chigangaidze et al., 2022). It expresses humanness in the values of compassion, sympathy, solidarity, sharing, harmony, consensus, and hospitality, which are crucial to designing (Mupedziswa et al., 2019). Ubuntu emphasises humanistic values. Empathy interviews are a tool that may prove helpful to designers if appropriately conducted from a humanistic perspective. Interviews in the inquiry phase help obtain contextual information about the problem, stakeholders, product or service use, and opinions about current products or services and gain expert insights (van Boeijen et al., 2013). It is ideal for conducting empathy interviews in the stakeholders' natural environment or problem context. Designers must understand their stakeholders' thoughts, feelings, motivations, and behavioural traits. This can be achieved through interviewing the stakeholders and, as such, helps designers to identify stakeholders' needs. It is essential to know the story of the stakeholders as it allows designers to connect with stakeholders, relate to the problem, and empathise. Using stories in the interview reveals personal insights and feelings that the designer can only be aware of by interacting with the potential stakeholder. This understanding helps designers innovate and create products or services that satisfy stakeholders' needs. Designers who actively listen to their stakeholders will feel valued and contribute more to their discussions.

Empathy interviews are essential to designers for the following reasons:

1 They allow stakeholders to speak about what is critical to their lives.
2 They centre on the expressive and intuitive aspects of the stakeholder.
3 They allow designers to gain insights into how stakeholders act in each context.
4 They expose solutions, and designers might not have discovered or overlooked them (Milton & Rodgers, 2013).
5 They assist designers in gaining deeper insights into the wicked problem.
6 They make the stakeholder feel at ease to speak from the heart.
7 They offer designers an opportunity to observe the body language and reactions of the stakeholders. This allows for spontaneous inquiries or probing based on observations (van Boeijen et al., 2013).

Empathy interviews are most appropriate when the developed products and services are familiar to stakeholders. Otherwise, context mapping and observations can be used (van Boeijen et al., 2013). Interviews eliminate the possibility of the interviewee misinterpreting the questions. Milton and Rodgers (2013) advance that there are three types of interviews: (a) unstructured – the designer asks stakeholders a series of open-ended questions, (b) semi-structured – the designer has a clear idea of what issues to be covered and ensures that central issues are covered, and (c) closed, fixed-response (structured) interview – all stakeholders are asked the same questions, and this type is helpful for novice interviewers.

The procedure for conducting empathy interviews can take the following order:

1 Make a list of questions to be asked.
2 Group the questions into themes or topics to make a smooth transition between the issues to enable the interview flows naturally.
4 At the beginning of the interview, introduce yourself and the project is undertaken.
4 Ask for consent to record the session.
5 Build rapport with the interviewee (ask for their name, where they come from, etc.).
6 Encourage stories.
7 Attempt to remain as neutral as possible and ask neutral questions.
8 Ask one question at a time.
9 Ask questions about the present prior to questions about the past or future.
10 Ask about specific instances or occurrences.
11 Look for self-contradictions; what people say and do can be very different.
12 Observe non-verbal cues, such as using hands, facial expressions, etc.
13 If you get stuck, ask "why?" Constantly asking why digs deeper into emotion and motivation. This helps the designer to understand stakeholder behaviour and identify needs.
14 Thank them and wrap up the session.
15 Devise a way to analyse the collected data and report the findings.

The limitation of empathy interviews is that the results cannot be generalised due to the small, selected interviewees (Ritchie & Lewis, 2003). However, the results can inform similar or related contexts (Flick, 2006). If the interviewer is not skilled in conducting interviews, this may affect the results, and stakeholders respond only to what they know consciously (van Boeijen et al., 2013).

Define Methods

These are methods used for synthesising the discovery phase to define the problem. As shown in Table 5.1, there are several ways to define the problem. However, this section will be limited to discussing the persona and mood board.

Persona

A persona is a detailed description of imaginary people constructed from well-understood, highly specified data about real people. A persona is a profile of typical stakeholders of the products, services, and system used to build a picture of the stakeholder, their preferences, characteristics, how they make decisions, etc. Cooper et al. (2007) argues that a persona is a stakeholder model representing specific characteristics of named individuals. A persona encapsulates a distinct set of behaviour patterns regarding using a particular product. Persona creation aims to identify target groups with similar objectives and expectations to a specific product or service. Personas help designers understand stakeholders, not as part of a group or a demographic but as individuals with a history, goals, interests, and a relationship to the product.

A persona is essential to design for the following reasons:

1   A product/service intended for a well-researched and well-specified target group has more chance of success in the market.
2   The best way to accommodate various stakeholders is to design for specific types of individuals with particular needs. When one broadly and arbitrarily extends a product's functionality, including many constituencies, one increases all stakeholders' cognitive load (Cooper et al., 2007). It may result in a product or service that may please some stakeholders and is likely to interfere with the satisfaction of others.
3   It helps the designer to understand stakeholders' motivations and needs and how these translate to a product or service use and, ultimately, purchasing decisions.
4   It increases engagement and communicates a shared vision among the design team.
5   Personas bring more excellent acuity, empathy, immediacy, and relevance to a new product or service development.

Despite the importance of personas in design, there are limitations to using the same such as:

1   They are a waste of time and resources if they are based on stakeholder data and are up to date because stakeholder behaviour is ever-changing.
2   They provide simulated and untrustworthy data, which can be a waste of time and resources.
3   It is not productive to generate models based on fictional characters. If they do not give the desired results, all the hard work is negative.
4   Most personas represent only a tiny portion of the potential stakeholder space.
5   It is hard for designers to empathise with a fictional character based on ambiguous and unvalidated research.

There are many ways of creating a persona. The best is to settle for a format that will provide quality and valuable information about the typical stakeholder. The following form is suggested for creating a persona.

1 Typical stakeholders – describe one or two specific stakeholders.
2 Pain points – what do stakeholders struggle with? What do they find frustrating?
3 Goals and expectations – what are the stakeholders' goals? What do they want to accomplish?
4 Task and activities – what specific tasks or activities do patrons do to achieve their goals?
5 Motivations, feelings, and skills – what motivates these tasks or roles? What are the stakeholders' attitudes and feelings towards these activities? What level of skills do they have to achieve their goals?

Figure 5.1 shows a persona presented in a graphic format.

Mood Board

A mood board is a physical or digital arrangement of images, materials, pieces of text, etc. intended to elicit or project a particular style or concept. It can include anything –such as photography, magazine cut-outs, designs or illustrations, colour palettes, textures, descriptive words –that helps one define the project's direction (Figure 5.2). Though the mood board in Figure 5.4 draws

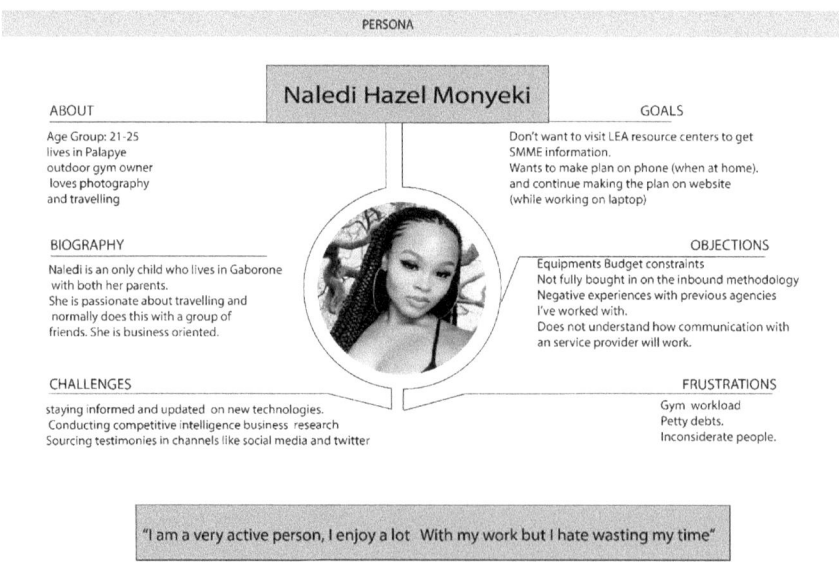

FIGURE 5.1 Persona

Source: Yaone Rapitsenyane

FIGURE 5.2 Sample mood board design on consumer lifestyle

Source: Bongani Sita

from the Afrikan culture, it also embraces what is globally available as part of design inspiration (think local and act global – glocalisation). The images may convey colour options, typographic layout possibilities, and the look and feel of the product, service, or system under development. Designers use a mood board to convey and inspire specific desires that facilitate creativity, innovative thinking, and application (Milton & Rodgers, 2013; Cassidy, 2011).

A mood board can communicate a stakeholder's subjective and emotional aspects of a design in the initial phases of the design process. To create suitable and effective boards, the designer should carefully consider the stakeholders' experiential characteristics and outstanding features in preparing mood boards. The mood board images facilitate interaction between the design and stakeholders at the early stages of the design process. Specific mood board types may include styling and theme boards, forecasting, colour, culture, material, and trend boards (Liu et al., 2020; Cassidy, 2011). In creating mood boards, the designer should use visual metaphors, e.g., a sports car may represent 'speed,' a lock communicates 'security,' vegetables may demonstrate 'greenness' or 'freshness,' a lion is synonymous with power – the king of the jungle. When using patterns and shapes in mood boards, the designer may use organic forms found in nature to reflect 'tranquillity' and 'calmness.'

In contrast, repeating patterns create visual 'energy' and 'movement.' After collecting all the needed materials for the mood board, the designer may build key themes using large images. Smaller images may be used around the big ones to support a key message being conveyed to stakeholders.

The mood board may be an impactful, memorable, and valuable tool to facilitate the creation of products, services, and systems. It also shows that it is effective, innovative, and fun to use, providing an effective technique for communicating the focus of the design (Reis & Merino, 2020). Mood boards can encourage the development of creative abilities, articulation, creative thinking, and can explore and communicate these visually. It assists the designer in exploring and experimenting with different media and visual presentation styles (Cassidy, 2011; Liu et al., 2020). Despite the advantages of using mood boards in design, there are several disadvantages: it can be overwhelming when dealing with much material to choose from, which can easily lead to plagiarising other people's work. Stakeholders often need to learn what an inspirational mood board is and its importance to product development. There are cases where stakeholders need to know what images to choose and may choose the wrong pictures, which will not inspire the creation of the desired product.

Business Model Canvas

The business model canvas assists designers in considering the economic relevance and context of the product or service being developed at an early stage when the design is still fluid. Designers begin by asking themselves questions, such as what value proposition the product or service brings to the market, who are the target stakeholders, how will the product or service generate revenue, what vital resources are needed to realise the product or service, who are the key partners in developing the same, etc. (IDEO, 2015). The business model canvas has nine segments (Table 5.6), as outlined by Osterwalder et al. (2004) and Osterwalder (2004). Table 5.6 shows an example of a sustainable business model canvas of vegetable delivery service by bicycle in all nine segments.

1 Key partners – Who are the key partners/suppliers? What are the motivations for the partnerships?
2 Key activities – What activities do the product/service value proposition require? What activities are the most important in distribution channels, customer relationships, and revenue streams?
3 Value prepositions – What core value does the product/service delivery to the stakeholder? Which stakeholder needs is the product/service satisfying?
4 Customer relationship – What relationship does the target stakeholder expect the designer to establish? How can that be integrated into the business in cost and format?

5 Customer segment – Which classes is the designer creating values for? Who is the essential stakeholder?

6 Key resources – What essential resources does the value proposition require? What resources are the most important in the distribution channels, customer relationships, and revenue stream?

7 Distribution channel – Which channels will be the most appropriate to reach stakeholders? Which channels work best? How much do they cost? How can they be integrated into the customers' routines?

8 Cost structure – What is more costly in the business? Which key resources/activities are the most expensive?

9 Revenue streams – At what value are stakeholders willing to pay? What and how have they recently paid? How would they prefer to pay? How much does every revenue stream contribute to the overall revenues?

TABLE 5.6 Vegetable delivery service by a bicycle business model canvas

Key Partners	Key Activities	Value Proposition	Customer Relationships	Customer Segments
* Mobile phone provider * Website provider * Bicycle factory sponsors	* Delivery of vegetables on a bicycle * Planning * Administration * Maintenance	* Eco-friendly, reliable, and cost-effective service * Express delivery * Fresh, organic vegetables	* Website * Face-to-face * Telephone * E-mail * Social media	* Anyone who needs fresh vegetables around the city
	Key Resources * Bicycles and helmets * Bags/containers to keep vegetables fresh * Laptops * Phones * Workers * Workspace		**Channels** * Pay online * Telephone * Word of mouth * Social media	

Cost Structure
* Workers   * Laptops
* Workspace   * Maintenance
* Insurance
* Website
* Smartphones

Revenue Streams
* Paying customers
* Corporate sponsorship

Some limitations of using a business canvas model include enabling risky assumptions within the business model without offering a straight-forward way to verify them. It focuses on the end outlook of the business without defining the business strategy to reach that milestone. It does not show any interconnections among the variables, as some variables will impact others.

Conceptualisation

This stage involves thinking up possible solutions that can address the problem. From Table 5.1, only the bisociation technique will be discussed to help Afrikan designers conceptualise innovative ideas based on their culture.

Bisociation

Bisociation is a lateral thinking technique based on the work of Arthur Koestler (1964). When two seemingly unconnected contexts or experience matrices are combined, they form a new relationship and develop shared meaning. The method involves bringing together two or more concepts whose connections were not previously suspected of creating a novel idea. The recognition of the two previously disconnected contexts or matrices is made compatible. They generate the experience of 'eureka,' or what Czarnocha (2014) referred to as 'an intuitive leap of insight' or an 'aha moment.' Blending different concepts results in combinatorial creativity – that is, new meaningful concepts and insights emerge through combining existing ones. Bisociation "uncovers, selects, re-shuffles, combines and synthesises already existing facts, ideas, faculties, (and) skills" (Koestler, 1964, p. 120). It plays a dual role as a cognitive re-organiser and effective liberator of a habit – it plants a double root for creativity (Czarnocha, 2014). According to Koestler, many bisociative creative innovations occur after a period of severe conscious effort directed at the creative goal or problem. It happens during a period of relaxation when rational thought has been abandoned. Figure 5.3 shows a design inspired by the bisociation technique based on Afrikan culture.

In Figure 5.3, the designer has associated two concepts: carrying the baby at the mother's back and jewellery design. Love provides an enduring bond between the baby and the mother. The designer drew an abstract shape of the baby at the mother's back and merged it with the heart shape to symbolise love. The new shape created is unique and provides a narrative of how babies were raised in Afrika. Unlike any other shape, the necklace piece conveys a unique narrative. People buy the narratives and fabulous experiences the design communicates. Afrika's culture inspired the design. Cultural design is a niche area Afrika can contribute to the design profession.

FIGURE 5.3   Necklace piece inspired by bisociation technique

Source: Caiphas Thusonyana Othomile

Prototyping

Prototypes are models built to test and evaluate a concept or process to act as an object to be replicated or learned from. It can be a working product made to exact specifications and used throughout the design development. It has functionality unlike a mock-up (minimal) or lack of it in aesthetic models. Designers use physical models to visualise information about the model's context. Prototyping is an effective way to make ideas tangible, learn through making, and quickly get critical feedback from the people they are designing for (IDEO, 2015). It is beneficial to test the concepts before production begins. Prototypes help the design team to discover issues related to manufacturing the final product. The design team also learns from the stakeholder through stakeholder feedback and stakeholder trials/interaction with the final prototype.

Prototypes can be made at varying levels of fidelity, targeting a range of stakeholders and environmental contexts. A combination of fidelity and stakeholder/environmental contexts allows for a deeper understanding of the ideas that aid design development. In Afrika, most students and designers use readily available sustainable resources for prototyping. Designers learn through making; fail fast and improve the design into the next iteration. The three levels of fidelity include:

1 Low fidelity – conceptual representation analogous to the idea; for example, paper prototypes enabling the stakeholder to offer input into the design idea (Figure 5.4).
2 Middle fidelity – represents features of the concept, such as a working mock-up with restricted functionality (Figure 5.5).

FIGURE 5.4   Quick and dirty low-fidelity prototype

Source: Photograph by the authors

FIGURE 5.5   Middle fidelity paper prototype of a crown

Source: Photograph by the author

3 High fidelity – mock-up of the concept, as feasible as possible to the final product, for example, a full-scale working prototype. It is tangible, testable, and allows entire stakeholder interaction on a mobile application and a web platform (Figure 5.6).

The details of the different types of prototypes will be discussed in Part 4 on prototyping approaches in this book.

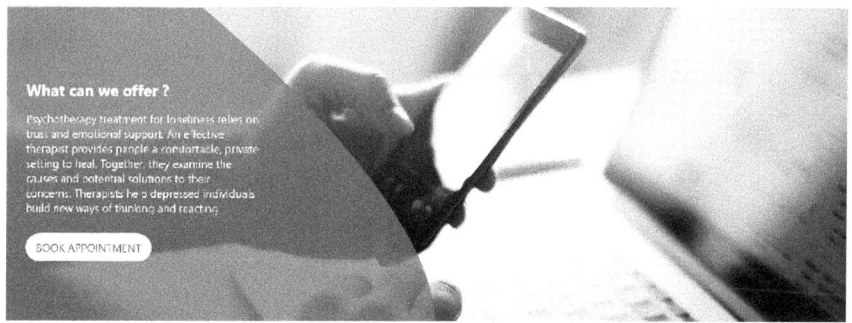

FIGURE 5.6   High-fidelity prototype

Source: Screenshot by the authors

Evaluation and Selection Method

The selection of high-quality ideas is a challenging task for designers. The aim is to select an original, desirable, feasible, and viable idea. Therefore, the datum method assists designers in selecting the most promising ideas that can be developed into complex concepts (Table 5.4).

Datum Method

Stuart Pugh (1991) developed the datum method or Pugh concept selection process. It is a method for the evaluation of design alternatives. One idea is to act as a datum to which the other alternatives will be compared against a range of criteria. Three judgements can be given: 'worse,' 'same,' or 'better' expressed in '–,' 'S,' and '+.'

The steps for using Pugh's method of concept selection (Pugh, 1991) include:

1 Choose the indicator by which the concepts will be evaluated – This criterion should include the stakeholders' needs and other design requirements identified during the problem definition stage. The least essential criteria should be removed if the initial list exceeds 15 standards to simplify the evaluation.
2 Formulate the decision matrix – Fill the row headings of the matrix, and the concepts fill the column headings.
3 Clarify the design concepts – This step aims to ensure that all team members understand each concept uniformly. This step assists in avoiding disputes and can lead to new and improved concepts.
4 Choose the initial datum concept – The datum is the concept with which all other concepts are compared. It is crucial to choose one of the better concepts as the datum. Selecting a poor datum will cause all the alternative concepts to look good and unnecessarily delay decision-making.
5 Run the matrix – During this step, each concept is compared to the datum for each criterion. Typically, a three-level scale is used: better (+), worse (–), or about the same (S). Better (++) and worse (––) are sometimes used.
6 Evaluate the ratings – Once the comparison matrix is completed, establish the sum of each concept's +, –, and S ratings. Refrain from allowing quantification to unnecessarily constrains the evaluation. If two or more top concepts are close, the strengths and weaknesses of each should be discussed. Be careful about rejecting concepts with highly negative scores. The few positive elements of these concepts can often enhance other concepts.
7 Establish a new datum and repeat the matrix – The new datum is usually the concept that received the most scores from round one. Using a new datum

to repeat the matrix delivers a different perspective for each comparison that will help clarify each concept's relative strengths and weaknesses. Ideally, the new datum will prove to be the superior concept. However, in some cases, there will be no clear winner. Regardless, the Pugh method aims to produce the most powerful concept or set of concepts for further development.

Table 5.7 shows that the C-Visor is the most promising idea because it satisfies seven criteria and has only two negative aspects compared to other ideas. In developing the C-Visor, positive features from the face guard can be borrowed to compensate for the two negative elements.

The matrix assists designers in finding the most promising or best design. It further prevents the design team from falling in love with a flawed design or meeting most design constraints. It is a communication tool and builds consensus based on the stakeholder's feedback. The results lead to significant cost savings.

Testing and Evaluation

In a competitive marketplace, testing and evaluation are critical activities for the success of any design to verify and validate its performance, reliability, safety, and durability. If a design is not tested, it can mismatch with stakeholders' needs, have technical design faults, or have issues regarding

TABLE 5.7 Sample Pugh selection matrix for a medical face shield

Selection Criteria	Design Concept			
	Armour	C-Visor	Face Guard	Eyeshield
Visibility		+	+	+
Durability		+	-	+
Comprehensive face protection	D	–	+	–
One size fits all	A	+	S	-
Easy to clean and reuse		S	+	+
Scratch resistance	T	–	+	+
Environmental impact		+	_	S
Manufacturing cost	U	+	+	–
Lightweight		+	-	+
Accommodate surgical mask and eyeglasses	M	+	S	+
+		7	5	6
S		1	2	1
–		2	3	3

manufacturability and maintainability (Tahera et al., 2012). Though testing can be expensive and time-consuming, it ensures product quality. A combination of virtual and physical testing is required to implement testing activities effectively. Virtual tests are done through computer-aided design modelling, while physical tests are conducted on prototypes. Designers often perform parallel and iterative product tests at several levels of product development, such as system, subsystem, and component. Although finding a problem at a system level may require a costly redesign of the product or service, this is expensive for the business. Therefore, it is advisable to conduct tests early in product development (Table 5.1). Some methods are used to measure the performance of a product or service, such as iterative testing, usability testing, co-discovery, task analysis, think-aloud protocol, etc. These methods assist the designer in analysing the design's strengths and weaknesses based on the stakeholders' feedback.

Iterative Testing

Iterative testing is a method in which a product or service is tested on stakeholders and constantly changed at various stages of the design process based on feedback. It involves gradually testing features as they are designed throughout the design process instead of waiting and testing them all at the end (Brown & Campbell, 2020; Hanington & Martin, 2017). It reduces usability issues before a product or service is launched. Iterative testing usually begins with low-fidelity prototypes (paper prototypes) and moves to middle-fidelity prototypes (working mock-ups with limited functionality) and high-fidelity prototypes (full-scale working prototypes). A good Afrikan example of iterative testing can be found in a Beegin beekeeping technology system developed by Brown and Campbell (2020). Klein (2013) advances that the advantages of identifying usability issues at the early stages of the design process help to shape a quality product or service.

Moreover, it is a cheap and easy way to collect feedback from stakeholders on the product's usability. This method facilitates the creation of efficient and streamlined designs with a better stakeholder experience because one can test, validate, and manipulate the design before it goes to the market. However, the disadvantages of this method are that more resources may be needed because of the frequency of stakeholder testing. The project may be prolonged due to testing results, making it hard to get to the market quickly.

Morales (2020) and TechSmith (2020) suggest that the following components facilitate iterative usability testing in products and services:

1 Learnability – How easily do stakeholders accomplish tasks the first time they encounter the design? Is the product or service easy to learn upon introduction?

2   Efficiency – As soon as stakeholders have learned the design, how quickly can they perform their tasks? Is there room for development in the proficiency of the task flows?

3   Memorability – Do stakeholders recall the interface when they return to it? Do stakeholders have to re-learn how to use the product or service after not using it? How easily can they re-establish proficiency?

4   Errors – How many errors do stakeholders make? Are they severe? Does the designer instruct them effectively in recovering from these errors?

5   Satisfaction – How satisfying is it to use the design? Are there features of delight and pleasant surprises for the stakeholder throughout their journey?

Concentrating on these five quality components during usability testing will enable one to identify any areas of product or service development, leading to a streamlined, efficient experience and quality design.

Usability Testing

It is a technique used to evaluate a product by testing it with the intended stakeholder group to improve the design, ease of use, and the stakeholder's perception of the experience (Niranjanamurthy et al., 2014). It evaluates task time, percentage of tasks completed, type and number of errors, and measures stakeholders' satisfaction. It focuses on measuring the design's ease of use and capacity to meet its intended function. The main intent of usability testing is to apprise the design process from the stakeholder's perspective; that is, the method verifies whether a design achieves its human interaction objectives. It involves representative stakeholders testing the design in a realistic environment. Once stakeholders start interacting with the design, a designer will observe how they use the product/service. The goal is to recognise usability problems, collect qualitative data, and determine the participants' satisfaction with the design (Almansour et al., 2020). By observing stakeholders' behaviour, emotions, and difficulties, designers can identify features that need improvement. Therefore, gathering and analysing verbal and non-verbal feedback helps the design team create a better stakeholder experience. Usability testing can be done in three ways:

1   Expert testing – Experts are asked to undertake usability testing of the design. They can benchmark the design concept against previous or competitor's products/services. The experts' in-depth understanding of the product type enables them to identify significant usability flaws early in the design process.

2   Formal stakeholder testing (laboratory usability testing) – In this method, the designer recruits about 5–10 test participants and gives them a set of scenarios that lead to the usage of specific features of a product (or

prototype). This method works best when one needs in-depth information on how real stakeholders interact with the product and their issues. It will help to investigate the reasoning behind specific stakeholder behaviour.

3 Informal stakeholder testing can be done in the format of guerrilla testing. With guerrilla testing, a product tester goes to a strategic place of interest, randomly selects participants, and asks them to play with or use a product for a given time and then provide feedback on the design.

Niranjanamurthy et al. (2014) and Almansour et al. (2020) suggest the following steps for usability testing:

1 Write down the usability test – creating tasks you want the stakeholder to execute.
2 Recruit various participants with various experiences that fit the target stakeholders.
3 Conduct a pilot test to validate the procedures and modify the necessary ones.
4 Perform the usability test – do not guide the stakeholder when performing the tasks. Questions should be asked after the test. Allow stakeholders to talk. The sessions should be recorded and notes should be taken. The results on a range of pass/fail should be recorded.
5 Evaluating the results – look for repeating patterns from the testing results.
6 Fix the identified problems and test again.

Some of the advantages of usability testing include: the designer can discover the actual demands and tasks of the stakeholder early in the design process, thus leading to cost-effective designs, and it decreases stakeholder confirmation time and errors, as well as stakeholder support, and provides a competitive benefit and satisfaction. Usability tests show better-quality products and services that are user-friendly. Stakeholders accept the products and services more readily, requiring less time to learn new features (Niranjanamurthy et al., 2014).

Conclusion

This chapter has provided a deeper insight into design research methods, their use, and possible tools. Samples of tools to use, how they have been used, and resultant outcomes have also been shared to provide more practical guidance on using this chapter with Chapter 3 on the new product development process. The design research methods and tools make the activities more organised, with room for documentation and reflection. The tools and procedures should not be prescriptive but more suggestive as the nature of design projects often differ, requiring different approaches and intuitive use of techniques at various stages of the development process. The central theme in design research is that stakeholders are the primary sources of information and inspiration since any development activities address their needs. Therefore, design research should not be about which tools to use but how to approach challenges at each stage of the project so that the development process does not become more of a tools and methods affair. Although the tools discussed in this chapter are not exhaustive, they can be adapted to developing products and services, especially in the early stages. It is crucial to contextualise the tools before using them in the Afrikan context. Otherwise, this may lead to design findings which are not authentic, and the resulting products and services will not solve society's challenges. The business case of any design project should also be investigated early, alongside the investigation of stakeholders' needs, hence the business model canvas. Further, the data analysis should ensure sensitivity to society's cultural norms and sustainability issues. The world now exists in a sustainability context, with everyone expected to ensure a sustainable future.

Reference List

Almansour, A., Osman, J. C., & Hamido, S. (2020). Laboratory-based usability test. *International Journal of Academic Scientific Research*, 8(2), pp. 1–15.

Barton, E. M., Baltes, M. M., & Orzech, M. J. (1980). Etiology of dependence in older nursing home residents during morning care: The role of staff behavior. *Journal of Personality and Social Psychology*, 38(3), 423–431. https://doi.org/10.1037/0022-3514.38.3.423.

Brand, K. G., & Campbell, A. D. (2014). In-context and ecology immersion for resilience: An exploration of the design of a household farming kit. *Proceedings of the 25th International Union of Architects World Congress: Architecture Otherwhere*, UIA 2014, pp. 1332–1343, Durban, South Africa. https://www.researchgate.net/publication/266146104_In-context_and_Ecology_Immersion_for_Resilience_An_Exploration_of_the_Design_of_a_Household_Farming_Kit.

Brown, I., & Campbell, A. (2020). Beegining: The implementation of appropriate technology. *Proceedings of the 9th International Conference on Appropriate Technology*, South Africa. www.researchgate.net/publication/346208797_Beegining_The_Implementation_of_Appropriate_Beekeeping_Technology.

Cassidy, T. (2011). The mood board process modeled and understood as a qualitative design research tool. *Fashion Practice*, 3, pp. 225–251. http://doi.org/10.2752/175693811X13080607764854.

Chigangaidze, R. K., Anesu Aggrey Matanga, A. A., & Katsuro, T. R. (2022). Ubuntu philosophy as a humanistic – Existential framework for the fight against the COVID-19 pandemic. *Journal of Humanistic Psychology*, 62(3), pp. 319–333. http://doi.org/10.1177/00221678211044554.

Chilisa, B., Major, T. E., Gaotlhobogwe, M., & Mokgolodi, H. (2016). Decolonising and indigenising evaluation practice in Africa: Toward African relational evaluation approaches. *Canadian Journal of Program Evaluation*, 30(3), pp. 347–362.

Cooper, A., Cronin, D., & Reimann, R. (2007). *About Face 3: The Essentials of Interaction Design*. Indianapolis: John Wiley & Son.

Czarnocha, B. (2014). Bisociation, the theory of aha! Moment. The basis of human mathematical creativity and for computer creativity of data mining. *Mathematics Teaching-Research Journal*, 17(1), pp. 1–18.

Daalhuizen, J. J. (2014). *Method Usage in Design: How Methods Function as Mental Tools for Designers. TU Delft*. Unpublished PhD thesis, Delft, the Netherlands: Delft University of Technology.

Dam, R. F., & Siang, T. Y. (2020). *How to Develop an Empathic Approach in Design Thinking*. www.interaction-design.org/literature/article/how-to-develop-an-empathic-approach-in-design-thinking.

Faste, T., & Faste, H. (2012). *Demystifying "Design Research": Design is Not Research; Research is Design*. Boston: IDSA Education Symposium. www.idsa.org/sites/default/files/Faste.pdf.

Flick, U. (2006). *An Introduction to Qualitative Research*, 3rd ed. London: SAGE Publications Ltd.

Hanington, B., & Martin, B. (2017). *Universal Methods of Design: 100 Ways to Research Complex Problems, Develop Innovative Ideas, and Design Effective Solutions*. Beverly: Rockport Publishers.

IDEO Organisation (2015). *The Field Guide to Human-Centred Design*. https://d1r3w4d5z5a88i.cloudfront.net/assets/guide/Field%20Guide%20to%20Human-Centered%20Design_IDEOorg_English-0f60d33bce6b870e7d80f9cc1642c8e7.pdf.

Klein, L. (2013). *UX for Lean Startups: Faster, Smarter User Experience Research and Design*. Sebastopol: O'Reilly Media Inc.

Koestler, A. (1964). *The Act of Creation*. Telford: Hutchinson.

Kolawole, E. (2015). Empathy. *IDEO Organisation. The Field Guide to Human-Centred Design*. https://d1r3w4d5z5a88i.cloudfront.net/assets/guide/Field%20Guide%20to%20Human-Centered%20Design_IDEOorg_English-0f60d33bce6b870e7d80f-9cc1642c8e7.pdf.

Liedtka, J. (2018). Why design thinking works. *Harvard Business Review*, September–October Issue. Retrieved from https://hbr.org/2018/09/why-design-thinking-works.

Liu, P. J., Chuang, M. C., & Hsu, C. C. (2020). Establishing picture databases for imageboards: An example for lifestyles of health and sustainability images. *Designs*, 4(21), pp. 1–17. http://doi.org/10.3390/designs4030021.

Martin, B., Hanington, B., & Hanington, B. M. (2012). *100 Ways to Research Complex Problems, Develop Innovative Ideas, and Design Effective Solutions*. Beverly: Rockport Publishers.

Milton, A., & Rodgers, P. (2013). *Research Methods for Product Design*. London: Laurence King Publishing Limited.

Moalosi, R., Marope, O., & Setlhatlhanyo, K. (2017). Decolonising Botswana's design education curricula by infusing indigenous knowledge: Botho co-creation process. In M. T. Gumbo & V. Msila (Eds.), *African Voices on Indigenisation of the Curriculum: Insight from Practice*. Wandsbeck: Reach Publishers.

Moalosi, R., Setlhatlhanyo, K., Sealetsa, O. J., & Marope, O. (2019). Co-created design solutions to mitigate human-elephant conflict in Botswana. In R. Halter & C. Walthard (Eds.), *Cultural Spaces and Design: Prospects of Design Education*. Basel: LIBRIUM Publishers.

Morales, J. (2020). *All You Need to Know about Iterative Usability Testing*. https://xd.adobe.com/ideas/process/user-testing/process-user-testing-iterative-usability-testing-best-practices/.

Mupedziswa, R., Rankopo, M., & Mwansa, L. (2019). Ubuntu as a Pan-African philosophical framework for social work in Africa. In J. M. Twikirize & H. Spitzer (Eds.), *Social Work Practice in Africa: Indigenous and Innovative Approaches*, pp. 21–38. Kampala: Fountain Publishers.

Nelson, H. G., & Stolterman, E. (2012). *The Design Way, Second Edition: Intentional Change in an Unpredictable World*. Cambridge, MA: MIT Press.

Niranjanamurthy, M., Nagaraj, A., Gattu, H., & Shetty, P. K. (2014). Research study on the importance of usability testing/User Experience (UX) Testing. *International Journal of Computer Science and Mobile Computing*, 3(10), pp. 78–85.

Norman, D. A., & Verganti, R. (2014). Incremental and radical innovation: Design research vs technology and meaning change. *Design Issues*, 30(1), pp. 78–96.

Olasunkanmi, A. (2015). Euthanasia and the experiences of the Yoruba people of Nigeria. *Ethics & Medicine*, 31(1), p. 31.

Osterwalder, A. (2004). *The Business Model Ontology – A Proposition in a Design Science Approach*. https://doc.rero.ch/record/4210/files/1_these_Osterwalder.pdf.

Osterwalder, A., Parent, C., & Pigneur, Y. (2004). *Setting Up an Ontology of Business Models*. CAiSE Workshops. https://ceur-ws.org/Vol-125/paper27.pdf.

Pugh, S. (1991). *Total Design: Integrated Methods for Successful Product Engineering*. New York: Addison-Wesley Publishing Company.

Rapitsenyane, Y. (2014). *Supporting SMEs Adoption of Sustainable Product-Service Systems: A Holistic Design-Led Framework for Creating Competitive Advantage*. PhD thesis, Leicestershire: Loughborough University.

Reis, M. R., & Merino, E. A. D. (2020). Mood boards: A systematic review of the literature on an imagery design tool focused on the aesthetic-symbolic definition of the product. *Estudos em Design*, 28(1). http://doi.org/10.35522/eed.v28i1.893.

Ritchie, J., & Lewis, J. (2003). *Qualitative Research Practice*. London: SAGE Publications.

Robson, C. (2011). *Real-World Research*. Chichester: Wiley.

Ross, J. (2018). *The Role of Observation in User Research*. www.uxmatters.com/mt/archives/2018/09/the-role-of-observation-in-user-research.php.

Sanders, E. B. N., & Stappers, P. J. (2014). Probes, toolkits, and prototypes: Three approaches to making in codesigning. *CoDesign*, 10(1), pp. 5–14. http://doi.org/10.1080/15710882.2014.888183.

Sauro, J. (2015). *4 Types of Observational Research*. https://measuringu.com/observation-role/.

Tahera, K., Earl, C., & Eckert, C. (2012). The role of testing in the engineering product development process. *Proceedings of TMCE 2012*, 7–11 May 2012, Open University, Karlsruhe.

TechSmith. (2020). *Usability Testing Basics*. http://webservices.itcs.umich.edu/drupal/wwwsig/sites/webservices.itcs.umich.edu.drupal.wwwsig/files/Usability-Testing-Basics.pdf.

van Boeijen, A. G. C., Daalhuizen, J. J., Zijlstra, J. J. M., & van der Schoor, R. S. A. (Eds.). (2013). *Delft Design Guide*. Amsterdam: BIS Publishers.

Winschiers-Theophilus, H., Bidwell, N. J., & Blake, E. (2012). Community consensus: Design beyond participation. *Design Issues*, 28(3), pp. 89–100.

Yu, E., & Sangiorgi, D. (2017). Service design as an approach to implement the value co-creation perspective in new service development. *Journal of Service Research*, 21(1), pp. 40–45.

6

ETHICS IN DESIGN

Richie Moalosi and Olefile Bethuel Molwane

Introduction

Designers need to consider the ethical consequences behind their decisions without being regulated. Technology is advancing rapidly, and regulating bodies cannot keep up with its pace. Often, regulations are made after the damage has already occurred. Therefore, designers must make ethical choices to solve society's challenges, take ethical responsibility, and regulate themselves by designing ethical products and services (Beard & Longstaff, 2018). Ethics empowers designers with knowledge about accepted norms and values associated with research activity. Ethics is a system of moral principles that define what is perceived as right and wrong (Falbe, 2018). It can also be described as a set of social rules, principles, and norms that guide the conduct of people in a society and as beliefs about right and wrong behaviour and good or bad character (Gyekye, 2011). Ethics assists in guiding designers about the dos and don'ts of design. It is an academic tool that assesses research quality, planning a research design, reporting, and publishing findings and results. Therefore, it instils scientific vigour in a study by preventing unethical misconduct. Ethical design is developed to design what is good for humanity or enhance the quality of life for people and for designers to take responsibility for their beliefs and actions (Beard & Longstaff, 2018; Leikas et al., 2019). Responsibility is mainly understood as socially, ethically, and environmentally acceptable activities for the upliftment of humankind. Therefore, this chapter discusses the ethical consequences of design from the perspective of Ubuntu philosophy. It also reviews the code of professional ethics and the benefits of ethical design to uphold the integrity of the design profession.

DOI: 10.4324/9781003270249-9

Afrikan Social Ethic

Gyekye (2011) argues that ethics is rooted in the patterns of behaviour that the members of the society consider bringing about social harmony and cooperative living, justice, and fairness. These are some of the ethos espoused in the philosophy of Ubuntu, which entails describing people's moral behaviour. Afrikan societies have undeniably evolved ethical systems – ethical values, rules – and principles that guide social and moral behaviour (Gyekye, 2011; Ogbechie & Anakwue, 2018). Though Afrika has diverse cultures due to ethnic pluralism, there are standard or shared features. For example, Afrikan society is closely related to the community and shared life of the people. The philosophy of Ubuntu assumes that there are certain fundamental norms and ideals which regulate people's conduct. For example, a person ought to conform to moral virtues that individuals can display in their conduct. The philosophy of Ubuntu's view is that what is good is constituted by deeds, behaviour patterns, and habits and considered by society as valuable due to their significance for human welfare (Gyekye, 2011; Udokang, 2014). Good deeds include such values as honesty, generosity, truthfulness, happiness, faithfulness, hospitality, compassion, peace, justice, respect, etc. that are meant to bring about social well-being. Actions that promote human welfare and social harmony are good and receive the community's approval, while the conduct that undermines human welfare and smooth social relationships is not approved. Thus, Afrikan ethics is a moral system based on humanistic ethics that focuses on human welfare, needs, interests, and social harmony (Appiah, 1998).

Therefore, designers from Afrika should express the perception of the society's common good, common humanity, and universal human welfare in their work. The common good encourages the creation of a social, moral, political, or legal system for improving the well-being of people in a community. It embraces the essential needs of all the community members and the enjoyment and fulfilment of the life of each member (Appiah, 1998). The common good is an essential facet of ethics embraced by the communitarian Afrikan society. This notion does not risk individual liberty as flaunted by Western liberal individualist thinkers. In this case, personal freedom is held as one of the essential goods of the members of society. Ethics ensures that individuals or groups participating in a study are treated with respect, confidentiality, human dignity, and free and informed consent.

Individualistic ethics that emphasises the welfare and interests of an individual is barely considered in Afrikan moral thought. Afrikan social ethic is expressed in many proverbs that underscore the significance of cooperation, mutual helpfulness, interdependence, collective responsibility, and reciprocal obligations. For example,

The well-being of man depends on his fellow man.
The right arm washes the left arm, and the left arm washes the right arm.

Your neighbour's situation is your situation.
A person is a person through other persons.

These proverbs call for the demonstration of interdependence, solidarity, sympathy, compassion, mutual help, goodwill, reciprocity, social well-being, cooperation, and willingness to offer each other some support. An individual cannot satisfy their life emotionally, socially, psychologically, economically, etc. without the assistance of other people. Ubuntu articulates the interconnectedness, common humanity, and the responsibility of individuals to each other. Therefore, this makes Afrikans a communitarian society, thus concerned for others and the community. Designers in the Afrikan context should respect these ethical considerations instead of being driven by economic considerations. It is essential to note that what is ethical does not always correspond with what is legal; for example, slavery was once legal but unethical (Baldini et al., 2018).

Ethical Design

Ethics in design guides how designers work with users, clients, and design team members, how they conduct the design process, how they determine the features of products and services, and how they assess the ethical-moral worth of the products and services resulting from the activity of designing. From an Afrikan perspective, designers should determine whether their practice will empower the community around them and develop the community. An ethical design must respect human rights and effort and enhance the human experience (Balkan & Kalbag, 2018). The design is ethical when it empowers users to frame their moral choices whenever value-laden issues emerge (Baldini et al., 2018).

There are several principles of ethical design, and these include:

1 Respect for human values – Designers should design products and services that account for human values (Balkan & Kalbag, 2018). They should consider the long-term systematic effects of current and future technologies on humans and the ecosystem.
2 Sustainability – Designers should strive to balance the socio-cultural, environmental, and economic developments by integrating ethical decision-making in inclusive and sustainable design practices. They should consider their work's impact on the environment, resources, and climate and embrace sustainability principles in the circular economy.
3 Usability – Designers are morally responsible for designing usable, pleasurable, and safe products.
4 User involvement – It makes sense to design with users because the design will become a part of their life. Their involvement will contribute to a positive experience.
5 Accessibility – Accessible designs benefit everyone, irrespective of their ability, disability, or age.

6 Respect for privacy – Design products and services that protect users' privacy and civil liberties (Leikas et al., 2019).

7 Transparency – Provide transparency so users can make informed choices about products and services. It relates to users having a right to know what they are signing up for when they agree to use a product or service or participate in a study.

8 Honesty – Design products and services that do not manipulate users with promises that cannot be fulfilled. Honesty ensures the design of trustworthy products and services.

9 Safety – It includes traits of the sanctity of life, inclusion, privacy, and emotional well-being.

10 Independence – It is crucial and refers to the user feeling in control of technology. If users are not in control of technology, it can be a frustrating experience, thus creating a sense of being a slave to technology.

Ethical design principles can be part of the code of conduct for designers. Figure 6.1 summarises the principles of ethical design.

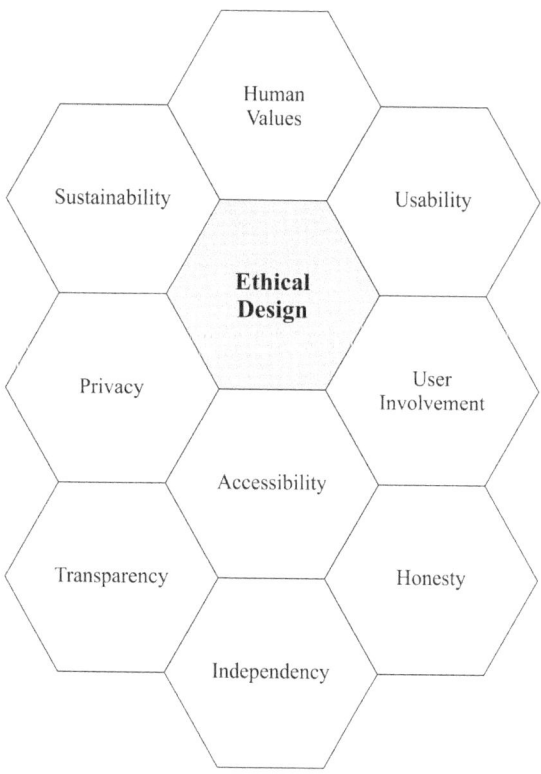

FIGURE 6.1 Principles of ethical design

Informed Consent

Design demands that the designer work collaboratively with users to understand their problems better. This compels the designer to act ethically and seek their informed consent, as it is necessary to adopt any design. Informed consent refers to gathering the user's agreement, encompassing criteria of disclosure and comprehension and voluntariness, competence, and agreement (Leikas et al., 2019). Users have the right to consent to any design or technological intervention. The goals of informed consent in research are to respect and promote a participant's autonomy and to protect participants from harm (Tam et al., 2015). Wilbanks (2018) argues that informed consent (i) provides an opportunity to inform research participants about clinical protocols; (ii) provides a meaningful choice architecture to support a potential participant's decision-making about whether to enrol; and (iii) builds an ongoing ethical relationship with study participants on trust-based permission.

To create an ethical design, the involvement of the users in the design phase should go beyond usability – it is about user experience design in the broadest sense, as it also entails experiences such as trust and comfort, including awareness of relevant value-laden choices (Baldini et al., 2018). Users must be informed about the risks, benefits, and alternatives of a given design intervention (Tam et al., 2015; Pols, 2017; Wilbanks, 2018). For example, users must be assured that their personal information will not be misused to expose them to risk. The users must be competent to make a voluntary decision on whether to participate in the project or not. Users who opt not to participate in the project should not be penalised for non-participation.

Tam et al. (2015) argues that when drawing informed consent, the following five concepts must be reflected: voluntariness, capacity, disclosure, understanding, and decision. On the same note, Friedman et al. (2005) advance six concepts: disclosure, comprehension, voluntariness, competence, agreement, and minimal distraction. Voluntariness means that an individual's decision to participate in the project is made without the investigator's coercion, intimidation, or persuasion. Capacity refers to the participant's ability to make resolutions that stem from understanding the information provided. Disclosure involves giving participants all relevant material about the project, including its nature, purpose, risks, potential benefits, and existing options. Understanding infers that participants can understand the information provided and appreciate its significance to their circumstances. The decision involves choosing to participate in the project or not. Minimal distraction refers to meeting the previous criteria of disclosure, comprehension, competence, voluntariness, and agreement without excessively diverting the individual from the task (Friedman et al., 2005). If the concepts are considered in informed consent, this helps to gain users' trust.

The designer should ask for participants' informed consent before the commencement of the research phase because it is mandatory to obtain it for all types of human subjects' research (Hayden et al., 2019; Hartson & Pyla, 2019).

The informed consent form should be written in a language easily understood by the users to minimise the possibility of coercion. Once users consent to participate in a project or study, they must sign an informed consent form indicating that they have read and understood the information provided, have answered all their questions, and decided to participate (Tam et al., 2015). The consent form is composed of an introduction, the purpose of the project, the type of research intervention, participant selection, the procedure of the project and its duration, risks and discomforts associated with participation (if any), benefits, issues of confidentiality, anonymity, assurance of their voluntary participation, and signatures – Table 6.1 (Hartson & Pyla, 2019). The form should also contain contact details other than the designer/investigator, whereby a participant can seek further clarity, including questions about the research, their rights as a research participant, or if they feel they have been mistreated (Hayden et al., 2019). If one is working with participants under 18, a parent or guardian must sign the informed consent form on behalf of the minor (Baxter et al., 2015). However, the investigator must verbally inform the minor of the same information to understand and consent. Finally, users should be given a copy of the signed consent form to keep for their records.

In the consent form, users should be assured of confidentiality issues, not disclosing information from the user, or identifying them using information they have provided (Tam et al., 2015). The participants' identities must be protected during the study, and their information should be kept confidential. Furthermore, participants should be assured of anonymity, that is, the protection against identification from the researcher. Pseudo names and codes must be used to ensure the anonymity and safety of all participants. All the records should have no real names of the participants to avoid associating any data with an individual (Tam et al., 2015; Hayden et al., 2019). It is crucial to assure participants that they will not incur any harm from the study. Harm during the study may be in the form of physical danger, stress, or loss of self-esteem. Participants should be informed that they shall be free to withdraw from the study at any given moment if they are no longer interested in participating without fear of being harmed or victimised (Hayden et al., 2019). Moreover, participants should be assured that the information they provide will be used for design purposes only and will not harm their private lives.

The final step of ethical clearance is to apply to an Ethics and Review Board within the university or relevant ministry in government. It is essential to check out the requirements that need to be completed by the Ethics and Review Board. In most cases, the applicant will attach the project proposal and relevant ethical clearance forms that must be filled out. Once the application satisfies all the Ethics and Review Board requirements, a research permit will be issued. The research permit will be used when recruiting participants. It gives them confidence that the project has been vetted by a competent regulating body and satisfies the required ethical standards.

TABLE 6.1 Sample consent form

[Your institution/organisation letter head]

Title of the project:
Name of principal investigator:
Project team member(s):
Name of organisation:
Name of sponsor:

This informed consent form has two parts:
• Information sheet (to share information about the study)
• Certificate of consent (signatures if you choose to participate)
You will be given a copy of the full Informed Consent Form for your records

PART I: Information Sheet
Introduction
Briefly state who you are and that you are inviting users to participate in a project which you are doing. Inform them that they may talk to anyone they feel comfortable talking with about the project and that they can take time to reflect on whether they want to participate or not. Assure the participant that if they do not understand some of the words or concepts used in this form, that you will take time to explain to them as you go along and that they can ask questions if they have any.

Purpose of the Research
Explain the research question(s) in simple terms which will clarify rather than confuse. Use local and simplified words rather than scientific terms and professional jargon. In your explanation, consider local beliefs and knowledge when deciding how best to provide the information. Investigators, however, need to be careful not to mislead participants, by suggesting research interests that they do not have. For example:

As part of the [project title], you are invited to participate in evaluating and improving various designs of [name of system or product], [description of system or product].

I give my consent (please tick all that apply):
• The researcher has explained the purpose of the research to me.
• I have had an opportunity to ask questions about the study.

Type of Research Intervention
Briefly state the type of design intervention that will be undertaken. This will be expanded upon in the procedures section, but it may be helpful and less confusing to the participant if they know from the very beginning whether, for example, the research involves an interview, observation, in-context immersion, focus groups, etc.

Participant Selection
Indicate why you have chosen this user to participate in this research. People wonder why they have been chosen and may be fearful, confused, or concerned.

(Continued)

TABLE 6.1 (Continued)

Voluntary Participation

Indicate clearly that they can choose to participate or not. State, only if it is applicable, that they will still receive all the services they usually do if they choose not to participate. It is important to state clearly at the beginning of the form that participation is voluntary so that the other information can be heard in this context. Although, if the interview or group discussion has already taken place, the person cannot stop participation but request that the information they provided should not be used in the research project. For example:

I give my consent (please tick all that apply):
- For the design team to observe me during the research.
- To audio record my voice.
- To video record my face.
- To record me when using a product or service.
- To note down my comments and actions.

Procedures

A Provide a brief introduction to the format of the research study.
B Explain the type of questions that the participants are likely to be asked in the focus group, interviews, or the survey. If the research involves questions or discussion which may be sensitive or potentially cause embarrassment, inform the participant of this.

Duration

Include a statement about the time commitments of the research for the participant, including both the duration of the research and follow-up, if relevant.

Risks

Explain and describe any risks that you anticipate or that are possible. The risks depend upon the nature and type of qualitative intervention and should be, as usual, tailored to the specific issue and situation.

Benefits

Benefits may be divided into benefits to the individual, community in which the individual resides, and society as a whole because of finding a solution to the research question(s) or design problem. Mention only those activities that will be benefits and not those to which they are entitled regardless of participation.

Confidentiality

Explain how the research team will maintain the confidentiality of data with respect to both information about the participant and information that the participant shares. For example, participants should be assured of issues of confidentiality, that is, not disclosing information from the participant or identifying them using information they have provided. Pseudo names and codes will be used to ensure the anonymity and safety of all participants. If the research is sensitive and/or involves participants who are highly vulnerable, explain to the participant any extra precautions you will take to ensure safety and anonymity. For example:

I give my consent (please tick all that apply):
- For people on the design team to play the recording in the future.
- For the researcher to include my anonymous comments in reports.

(*Continued*)

TABLE 6.1 (Continued)

Sharing the Findings

Your plan for sharing the findings with the participants should be provided. If you have a plan and a timeline for the sharing of information, include the details. You may also inform the participant that the research findings will be shared more broadly, for example, through publications and conferences, seminars, workshops, and meetings.

Right to Refuse or Withdraw

This is a reconfirmation that participation is voluntary and includes the right to withdraw. Ensure that the section fits the group for whom you are seeking consent. Participants should have an opportunity to review their remarks in individual interviews and erase part or all the recording or note.
• I understand that I can leave at any time without giving a reason.

Who to Contact?

Provide the name and contact information of someone who is involved, informed, and accessible – a local person who can be contacted. State also the name (and contact details) of the local Institutional Review Board that has approved the proposal.

This proposal has been reviewed and approved by [name of the local Institutional Review Board], which is a committee whose task it is to make sure that research participants are protected from harm. If you wish to find out more about the local Institutional Review Board, contact [name, address, and telephone number]. It has also been reviewed by the Institutional Review Board of the (XX University), which is funding/sponsoring/supporting the study.

You can ask me more questions about any part of the research project if you wish to. Do you have any questions?

PART II: Certificate of Consent

This section must be written in the first person. It should include a few brief statements about the research. If the participant is illiterate but gives oral consent, a witness must sign. A researcher or the person going over the informed consent must sign each consent. Because the certificate is an integral part of the informed consent and not a stand-alone document, the layout or design of the form should reflect this. The certificate of consent should avoid statements that have "I understand. . . ." phrases. The understanding should perhaps be better tested through targeted questions during the reading of the information sheet or through the questions being asked at the end of the reading of the information sheet, if the potential participant is reading the information sheet him/herself.

I voluntarily agree to participate in this study, and I know of no reason I cannot participate. I have read and understand the informed consent and conditions of this project. I have had all my questions answered. I hereby acknowledge the above and give my voluntary consent for participation in this project. If I participate, I may withdraw at any time without penalty. I agree to abide by the rules of this project.

(Continued)

the fees, including all necessary conditions. The contract should also be based on the code of ethics requirements and respect the fundamental laws establishing the designer/client or employer relationship.

2 Designers should not knowingly infringe on the intellectual property rights of other designers.

3 Designers should not engage in any fraudulent activity in their practice.

4 Designers are expected to treat all knowledge and information about their client's or employer's business as confidential. Such information shall not be divulged to any third parties without the consent of the relevant client or employer.

5 A designer should not knowingly assume or accept a position where personal interests conflict with professional duty.

6 Design professionals should uphold human rights in all their professional undertakings. For example, a designer should refuse to engage in or support discrimination against individuals based on race, gender, age, religion, ethnic origin, national origin, sexual orientation, or disability.

7 A designer should neither offer nor make any payment or gift to a public official to influence the official's judgment in connection with an existing or prospective project in which the designer is interested.

8 A designer serving in a public capacity shall not accept payments or gifts intended to influence their judgment.

9 A design professional should not alter the contractually defined scope, objectives, or catalogue of deliverable components of a project without the client's consent.

10 Designers should avoid conflicts of interest in their professional practices and fully disclose all unavoidable disputes as they arise.

11 Designers should safeguard the trust placed in them by their clients and users.

12 Design professionals should employ reasonable safeguards in systems and practices to protect the confidentiality of hard copies and digital project materials.

13 Designers should uncompromisingly maintain their honesty, integrity, and transparency, thus avoiding deceptive acts.

14 Designers should avoid all conduct or practice that deceives the public.

15 Designers should not attempt to obtain employment, advancement, or professional engagements by untruthfully criticising other designers or other improper or questionable methods.

16 A designer should give credit for design work to those to whom honour is due and should recognise the proprietary interests of others.

17 A designer should not undertake any work at the client's invitation without paying an appropriate fee. However, a designer may undertake work without a cost or at a reduced rate for charitable organisations.

Benefits of Ethical Design

The ethical design has several benefits, such as:

1 It reduces business risks for investments in products and services from a legal perspective.
2 It supports businesses in a long-term relationship with users who want to buy ethically framed products and services and are willing to use those that better meet their needs.
3 It helps create a society where users feel good about using products and services with a relatively high level of trust (Baldini et al., 2018).
4 It produces societally acceptable and desirable products and services that avoid unintended consequences and manage commercial risks (Leikas et al., 2019).

Conclusion

Ethics can be considered an academic tool that can evaluate the quality of research in terms of planning a research design, reporting, and publishing findings and results. It is embedded in society's ideas and beliefs of what is good and evil. What is good is considered acceptable by society, and what is evil is viewed as anti-social and unacceptable behaviour. Ethical values, principles, and rules are meant to guide social and moral behaviour in the community, and the same is true for the design profession. Ethics also advance social harmony, justice, fairness, and cooperative living, as espoused in the philosophy of Ubuntu. Afrikan ethical considerations stem from communal traditions (not based on individual consent), anti-universalising, humanist, and naturalistic. Ubuntu places a critical value on the human being (human-centred) and the community (community-centred).

Society expects a designer to develop ethical design business practices and should responsibly conduct their business. In Afrikan design, business practices should be influenced by indigenous knowledge, technologies, values, and trends due to the cultural diversity of the people, places, and values. It is significant for designers to understand and situate indigenous ethical principles and practices in Afrika within the design business framework. This chapter has critically discussed ethics in design from an Afrikan perspective. Designers should understand that Afrikan ethics is a moral system based on humanistic ethics that focuses on human welfare, needs, interests, and social harmony. Designers should ensure that ethical principles that helped to maintain social order and stability in traditional Afrikan societies should be upheld for continuity. They should design products and services that display Ubuntu ethos, such as promoting communal relationships with other people. This chapter discusses ethics as a philosophical unity of values, ideals, character, and collaborative relationships with other people. Designers need a deeper understanding of Afrikan ethics to make ethical choices that advance society's human welfare and social relations. Ethics must include critical and relevant social values. The chapter concludes by advocating for a healthy synchronisation of the positive elements of traditional Afrikan Ubuntu ethics with Western culture for a better moral society in modern Afrika, especially in designing ethical products and services.

Reference List

Appiah, K. (1998). Ethical systems, African. In *The Routledge Encyclopedia of Philosophy*. Taylor and Francis. www.rep.routledge.com/articles/thematic/ethical-systems-african/v-1. http://doi.org/10.4324/9780415249126-Z008-1.

Baldini, G., Botterman, M., Neisse, R., & Tallacchini, M. (2018). Ethical design in the internet of things. *Science Engineering Ethics*, 24, pp. 905–925.

Balkan, A., & Kalbag, L. (2018). *Ethical Design Manifesto*. https://2017.ind.ie/ethical-design/.

Baxter, K., Courage, C., & Caine, K. (2015). *Understanding Your Users: A Practical Guide to User Research Methods*. Amsterdam: Elsevier Inc.

Beard, M., & Longstaff, S. A. (2018). *Ethical Principles for Technology*. Sydney: The Ethics Centre.

Falbe, T. (2018). *Ethical Design: The Practical Getting-Started Guide*. www.smashing-magazine.com/2018/03/ethical-design-practical-getting-started-guide/.

Friedman, B., Lin, P., & Miller, J. K. (2005). Informed consent by design. In L. Cranor & S. Garfinkel (Eds.), *Security and Usability*, pp. 495–521. Sebastopo: O'Reilly Media.

Gyekye, K. (2011). *African Ethics, The Stanford Encyclopedia of Philosophy*, Fall 2011 ed. (E. N. Zalta, ed.). https://plato.stanford.edu/archives/fall2011/entries/atrican-ethics/.

Hartson, R., & Pyla, P. (2019). *Empirical UX Evaluation: Preparation in the UX Book*, 2nd ed. Amsterdam: Elsevier Inc.

Hayden, D. M., Saclarides, T. J., & Bhama, A. R. (2019). Is the informant adequately informed? Attitudes regarding the informed consent discussion. *Journal of American College Surgeons*, 229(4), p. E187. http://doi.org/10.1016/j.jamcollsurg.2019.08.1241.

Leikas, J., Koivisto, R., & Gotcheva, N. (2019). Ethical framework for designing autonomous intelligent systems. *Journal of Open Innovation: Technology, Market, and Complexity*, 5(1), p. 18. https://doi.org/10.3390/joitmc5010018.

Lucivero, F. (2016). *Ethical Assessments of Emerging Technologies: Appraising the Moral Plausibility of Technological Visions*. Heidelberg: Springer.

Ogbechie, R., & Anakwue, N. (2018). Ethical principles and practices in Africa, indigenous management practices in Africa. *Advanced Series in Management*, 20, pp. 205–219. https://doi.org/10.1108/S1877-636120180000020011.

Pols, A. J. K. (2017). May Stakeholders be involved in design without informed consent? The case of hidden design. *Science Engineering Ethics*, 23, pp. 723–742. https://doi.org/10.1007/s11948-016-9811-0.

Rams, D. (2011). *Less and More: The Design Ethos of Dieter Rams*. Berlin: Gestalten.

Tam, N. T., Nguyen Tien Huy, N. T., Thoa, L. T. B., Long, N. P., Nguyen Thi Huyen Trang, N. T. H., Hirayama, K., & Karbwang, J. (2015). Participants' understanding of informed consent in clinical trials over three decades: Systematic review and meta-analysis. *Bulletin of the World Health Organization 2015*, 93, pp. 186–198. http://doi.org/10.2471/BLT.14.141390.

Udokang, E. J. (2014). Traditional ethics and social order: A study in African philosophy. *Cross-Cultural Communication*, 10(6), pp. 266–270. http://doi.org/10.3968/5105.

Wilbanks, J. (2018). Design issues in E-consent. *The Journal of Law, Medicine & Ethics*, 46(1), pp. 110–118.

World Design Organisation. (2020). *Code of Professional Ethics*. http://uploads.wdo.org.s3.amazonaws.com/ProfessionalPractice/WDO_CodeofEthics.pdf.

PART 3

Design Cognition

7

DESIGN ELEMENTS AND PRINCIPLES

Shorn Molokwane, Odireleng Marope, and Matthews Ollyn

7.0 Introduction

Design has guiding tenets to which all design practice refers for guidance in the execution or arrangement of objects into compositions (McClurg-Genevese, 2005). Design elements and principles are the fundamental building blocks of any composition, especially in the visual and applied arts field. These building blocks are arranged to form different comprehensive compositions, which, when combined well, create product outcomes that effectively communicate with the end user. Elements and principles of design are applied in products and services such as scenic design or set design (Brako & Gilbert, 2022). They are popularly applied in the aesthetics of various products, including building integrated photovoltaics (BIVs) for adoption by homeowners and architects (Awuku et al., 2021). In the case of Brako and Gilbert (2022), elements and principles have demonstrably been used in TV3's *Ghana's Most Beautiful* reality show (2017), where the mood and themes for the designed show were defined and attracted a good number of audience and positive emotions from watching the show.

Elements considered for design differ with the varying literature (Adams, 2013), and how the writers interpret the design discipline. Gatto et al. (2011) define visual design elements as ingredients designers can use. According to them, these include line, colour, shape, form, value, space, and texture. Evans and Thomas (2012) identify the elements as line, texture, shape, space, type, size, colour, value, and volume. Designers, artists, and other creative people combine these elements into variegated arrangements in principles of design. These principles can be applied to products across all design disciplines, such as architecture, art, product design, graphic design, interiors, fashion, etc.

DOI: 10.4324/9781003270249-11

The principles of design combine the design elements to give form and coherence, which can be interpreted by the designer or creator and used in combination with other elements to create a product or service that has aesthetic significance and evokes desired user experience. The principles form the basis for the characteristics of the product and service. Soğukkuyu (2021) defines design principles as a set of rules that consists of balance, integrity, proportion, rhythm, emphasis, contrast, and movement elements that allow the design to be systematically formed. These different combinations form the basis for the vast variances of creativity available today. In works of art, as an example, the design elements would be combined in an arrangement that follows basic principles and considerations in a combination known as composition.

Design elements and principles have primarily been treated from a Western/Eurocentric perspective. The absence of other perspectives may render these understandings incomplete when applied broadly. Afrikan design is a diverse and rich field that draws inspiration from various sources, including cultural traditions, natural materials, flora, fauna, and contemporary aesthetics. The elements and principles of Afrikan design reflect the continent's unique cultural and artistic traditions.

This chapter discusses the elements and principles of Afrocentric and Eurocentric elements and principles of design. Moreover, the chapter discusses the Western influence on Afrikan design and the Afrikan interpretation of elements and principles of design in Afrikan aesthetics. The Afrikan cultural symbols and artefacts were reviewed. A case study has been reviewed on the application of Afrikan elements and principles in design. The chapter concludes by harmonising Afrocentric and Eurocentric interpretations of elements and principles of design.

7.1 Elements of Afrikan Design

The elements of Afrikan design include some of the following:

1 Colour: Afrikan design is known for its bold and vibrant colour palettes that reflect the richness of Afrikan cultures and traditions. Colours such as red, yellow, green, and blue are often used to create striking visual contrasts.
2 Texture: Texture is an essential element of Afrikan design, with many designs incorporating natural materials like wood, stone, and clay. Textiles are also an essential element of Afrikan design, with many designs featuring intricate weaves, patterns, and embroidery.
3 Form: Afrikan design often emphasises organic, flowing forms that reflect the natural world. This can be seen in the use of curvilinear shapes and motifs and in the incorporation of natural elements such as plants and animals.

4 Symbolism: Afrikan design frequently incorporates symbols and iconography that are meaningful to specific cultures and communities. These symbols can represent everything from spiritual beliefs to social status.

7.2   Principles of Afrikan Design

The principles of Afrikan design include some of the following:

1 Unity: Afrikan design often emphasises unity and community, with designs that bring people together and celebrate shared traditions and values.
2 Harmony: Harmony is an essential principle of Afrikan design, with many designs seeking to create a sense of balance and coherence using colour, form, and texture.
3 Balance: Afrikan design often incorporates elements of balance and symmetry with visually balanced and harmonious designs.
4 Proportion: Proportion is an essential principle of Afrikan design, with designs that use scale and proportion to create visual interest and balance.
5 Emphasis: Afrikan design often uses emphasis to draw attention to specific elements or motifs, creating a focal point that draws the viewer's eye.
6 Movement: Movement is an essential principle of Afrikan design, with many designs incorporating flowing lines and shapes that suggest movement and energy.

Overall, Afrikan design is a rich and complex field that incorporates many elements and principles to create visually striking designs and culturally meaningful and artistically compelling designs.

7.3   Eurocentric Elements and Principles of Design

The elements and principles of design are first addressed from a Eurocentric perspective because it is a standard used globally. Eurocentric elements of design include the following:

* Line – a continuous mark connecting two points. It may be actual, implied, vertical, horizontal, diagonal, and/or contour.
* Shape – a two-dimensional element with the area on a plane surface.
* Form – a three-dimensional element with volume in space.
* Size – a measure of the smallness or bigness of the element.
* Space – the distance or area around or between elements in creation.
* Colour – the visible spectrum of radiation reflected from an object.
* Value – how light or dark an object, area, or element is, independent of its colour.
* Texture – the surface quality of a shape or how it appears: rough, smooth, spiky, soft, hard, glossy, etc.

Once organised into matrices and combinations, the design elements would lend themselves to design characteristics such as harmony, symmetry, movement, rhythm, etc. It is from these characteristics that the object–user relationship is sought and developed.

In his book *Introduction to Creative Design*, Edel (1967) outlines the aspects the products may have to be considered attractive as: rhythm, dominance, balance, transition, variety, contrast, and unity.

1 Rhythm is an aspect of the object having a 'beat', i.e., the arrangement of the elements in a way that are repetitive in regular intervals. The rhythm may be visual, audio, tactile, or many other forms.
2 Emphasis/dominance says some aspect of the object is accentuated by concentrating some design elements or combining them to give the object its character through the highlighted part. This may be done, for example, by weight distribution, colours, scaling, etc.
3 Balance addresses combining or arranging the design elements to create a sense or feeling of stability of the object; this may be achieved in various ways, for example, in symmetry or weight balancing, colour arrangements or visual arrangement, material combination, and distribution. Balance is closely related to unity.
4 Unity is the quality yielded by the object in the effectiveness with which all the elements are combined to make a product, which is a whole unified piece in concept and effect. It is closely related to harmony, which is about harnessing similar characteristics of the elements to be represented.
5 Transition/Movement is a basic motion or motion depiction achieved by arranging the design elements, for example, a trophy of a sculptured sports player in action.
6 Harmony combines all the design elements with congruence with one another and with the whole or creating a family set of products. Similarities and differences in the elements are used to create an 'agreement' in the combinations.
7 Proportion is the relation of the elements to the whole object and how they relate. Depending on the product's functional or decorative requirements, some elements may be emphasised.
8 Variety adds a difference to the work when different elements are put together.
9 Contrast is a unique form of variety when juxtaposing opposing elements produces an interesting effect.

7.4 Design in Afrika Prior to Western Influence

Design had existed for centuries in Afrika, long before the advent of the colonisation of Europeans, whose presence catapulted the design drive in a different direction. It is worth noting that design has evolved over the period with evolving lifestyles, needs, demands, and economies of the people (both makers and consumers) involved. Natural resources and available materials also influenced the design's form and nature. The different activities and processes in the built and construction environments, mining and metallurgical works which have been going on around the Afrikan continent since some 2000 years ago indicate high levels of civilisation which were achieved through designed products, skills, and knowledge. Examples include impressive architectural works, from the pyramids in Egypt and the 13th-century incredible cities in the empires of Timbuktu in Mali, and massive and impressive structures of the 11th century of the great wall of Zimbabwe.

Moreover, the variegated arts and crafts activities demonstrated a way of creating and doing, which could only be achieved using tools and implements, hence the existence of design before the advent of Europeans in Afrika. Van Sertima (1983) and Brooks (1971) speak of making tools and implements such as metal chisels, saws, copper and iron tools, weapons, nails, glue, artefacts, art, and even steam engines across the Afrikan continent during those times. Many other design activities included extensive textiles, ceramics, woodcraft, and metalsmith work throughout the Afrikan continent. The examples depicted deep cultural connotations and were often commissioned by tribal leaderships to celebrate their variegated activities or rituals, such as the kente textile designs in Ghana.

The Industrial Revolution influenced the development of industrial design in Europe, e.g., the agricultural pursuits, exploration of minerals, and trade necessities between nations shaped the evolvement of industrial designs across the continent. The designs of most objects were strongly utilitarian, often adorned with details of cultural contexts and significance. This would depict origin, maker, period, status, intent, value, etc. Over the years, as Afrikan design work and practice interacted with other cultures from beyond the continental shores, there was a gradual integration and assimilation of especially Western influences, which in part contributed to what is currently regarded as contemporary Afrikan design.

It is commonly agreed that Afrika is a continent of great diversity, comprising 54 countries with their own diverse cultures, and at the same time having a commonality of Afrocentricity. Often, design works from Afrika have suffered stereotypical interpretations and classification from Western perspectives, which has yet to help its advancement. This is being addressed by Afrikan scholars, designers, and other stakeholders at an individual and continental level. Different interest groups converge on initiatives to inform, showcase, educate, and promote Afrikan design in its true diversity, commonality, and role. These include, amongst many others, the Pan Afrikan Design Institute, Design Afrika, Network of Afrikan Designers, Design Indaba Expo, Design Network Afrika, Design Weeks hosted by different countries, fashion shows, design conferences, seminars and workshops, and collaborative design activities across the continent.

7.5 Western Influence on Afrikan Design Commence

To understand the history of Western influence on Afrikan design, we must relate to the influence of Western culture on Afrikan cultures, which can be traced to the colonial period and Afrika's initial contact between these cultures. The Western influence came through the European colonists, the English, French, Portuguese, Germans, Dutch, and Belgians from the 1600s. The Europeans' lifestyle manifested through the tools and implements they used per the context. Afrikans created a visual reference to adopting/adapting some of these naturally into their lives over the centuries.

The Western and Afrikan tools and implements differed in many aspects. The difference was in materials and manufacturing processes, even those with similar utilitarian intentions. Transitions such as the Industrial Revolution in Western Europe would have seen the development of tools, equipment, and industrial processes more evolved than what the Afrikans used at the time, the latter not having had the impact of World War II, which triggered such a revolution. There were many factors for the spread of Western influence amongst the Afrikans. This included the perceived superiority of Western culture over Afrikan cultures. Some local groups yearned to belong to what they believed to be a better way of living.

7.6 Afrikan Interpretation of Elements and Principles of Design in Afrikan Aesthetics

The approach in this chapter introduces a paradigm shift in using Afrikan cultural and contextual references. In "The Afrikan Aesthetic as It Informs Product Form," Molokwane (2007) discusses some of the typical elements and principles of design from an Afrikan perspective. These principles are discussed elsewhere by Okpewho (1977), Evans and Thomas (2012), Yeo and Cao (2021), and Vogel (1986), basically stressing their importance and relevance in the global design discourse. These would vary in emphasis across the many different cultures constituting the Afrikan continent; hence a holistic approach is adopted. Whilst not exhaustive, scholars of Afrikan creativity concur that principles and elements of Afrikan aesthetics include togetherness, craftsmanship, human physical and spiritual symbolism, religion, oral traditions, self-composure, a close relation to nature, empathy, simplicity and elegance, form-giving, liberty, and humility (Molokwane, 2007). The following section will discuss some of the elements and principles of Afrikan aesthetics influenced by the philosophy of Ubuntu.

7.6.1   *Religion*

Religion is an essential part of any society, and it varies in form and practice, just as people vary in race and ethnic structures. Most ethnic groups, whether separately or collectively, believe in a superior being, usually one, who is thought to be the creator of all that is natural around, the provider and guidance of life. There is collective worship, song, dance, and offerings, thanksgiving to the creator, asking for continual providence and guidance of the essentials of the good life. Sometimes there are several gods to whom the people pray, representing different aspects of the society, such as fertility, rain, health, etc., unlike the god structures of the ancient Greeks (Figure 7.3). The difference is that in most cases, there is one ultimate God, similar in the ultimate singularity of a supreme being like in Christianity, Islam, or any other religion with a unique God to whom all worship.

7.6.2   *Oral Tradition*

Oral traditions are a central part of the Afrikan way of life, practised in spoken word, songs, chants, and other forms across the continent. Proverbs, sayings, dictums, verses, and many other poetic formulations are loaded with meaning, reflecting society's outlook and experience in all aspects of life. This then translates to the Afrikan aesthetic outlook that is directly and indirectly expressed in various forms of creative pursuits, which, according to Abusabib (1995), typically comprise deep artistic and aesthetic insights, reflecting the existent principles and loaded with creativity, criticism, evaluation, etc.

Abusabib (1995) addresses the difficulties in approaching oral traditions as a source of data on creative activities, arguing that people with first-hand exposure and experience of these cultures can only unpack the details of these messages and go deeper into their meaning. This necessarily places those intimated with the cultures and indicative practices best positioned to inform the creative discourse and related activities, arguing that it is incumbent on the Afrikan designers and creators to champion the role and place of Afrikan design on the global map.

7.6.3   *Form-Giving*

The other aspect of the Afrikan aesthetic manifests itself well through form-giving practices, especially handcrafts and the arts – form-giving assists in elaborating the objects. There are several stages involved. Usually, the workers would be talented artisans who would have learnt and perfected the skills over long periods, typically passed on from generation to generation. There is great pride in the creator's skills, and the product end users have a special relationship or connectedness with the artisan. In some cases, it is not uncommon for a master artisan to be accorded a special place in the hierarchy of social leadership because he is intricately involved in shaping people's lives; he lives through the objects he creates for them. The object creation process starts with the clients' all-important needs and the provision of services for their everyday needs. The raw materials are then sourced from nature, using sustainable extraction and consequential processing methods that cause minimum disturbance to the ecosystem.

The development process can be followed should the user wish to see it, as the artisan usually works from the same communal setting. The product or object will necessarily incorporate the user's cultural context, as the artisan is well vested in such knowledge and recognises the importance of the object speaking the user's language. The language may be reflected in shaping the object, colouring, texture, decoration, or smell. These and similar factors such as size, weight, etc., essentially are the design elements that a designer or artisan would consider in product development. Different combinations of these would result in various objects, which would consequently be received with various reactions according to the context of the application and the user's circumstances.

In most objects, design and decoration, semiotics, which are about symbolism and meaning, play an essential part. The symbols and signs infused into the different design elements give rise to exciting interpretations, understanding and aesthetic appreciation of the objects. Artisans encode different interpretations of people's lives, cultures, myths, beliefs, etc., in colours, shapes, textures, etc. The elements and principles used may speak to different aspects of the user's life, such as health, social prestige, wealth, power, etc. Figure 7.1 is a bowl made of wire and decorative beads taken from the Zulu culture to protect them against evil spirits.

7.6.4 Togetherness

Afrikans love doing things in unison; for example, in most family chores, the mothers will be working on household activities with their daughters, and the fathers and the uncles with the sons and nephews. For most tribes living in rural settings, the culture of gathering at night around a campfire, eating and telling stories, still forms a central part of family reunions. The bond and passion for sharing are evident in traditional rituals, such as weddings, funerals, and cultural celebrations, where close family members and friends would help in the preparations of the ceremony. This may be in food preparations, set-up and decorations, or other organisational chores.

7.6.5 Craftsmanship

Most objects are handmade by a group or family with a long craft tradition, skills, and knowledge. The objects are usually made intricately, with exquisite

FIGURE 7.1   A beaded wire fruit bowl

Source: Photograph by the authors

details, and to an excellent finish. Typical items would include baskets, crockery, furniture, cutlery, etc. Ornamental pieces such as sculptures and paintings are also popular. The objects typically have symbols or signs in their design, carrying a meaning or contextual significance. The basket patterns in Figure 7.2 depict a running ostrich. This interpretation is drawn from the people's interactions with animals and nature and bear significance to their lives. The owner/user of such an object may expect a particular impact on their life, depending on their cultural understanding and belief related to interpretations of the designs.

FIGURE 7.2 Traditional woven basket depicts various Afrikan elements and princi-ples, such as rotational symmetry, alternating rhythm, texture, colour, and myth and meaning embedded in the shapes

Source: Photograph by the authors

In the design of baskets, for example, there are two primary considerations: the utilitarian nature of the object and the decorative end. Foodstuff or other household items, such as crockery and laundry baskets, are usually designed in a typical bowl and bucket format, semi-spherical or semi-spheroidal. Such forms are chosen for their practicality in containment, the easiness of shaping the object and the harmonious effect that the curvaceous detail gives to the finished object. When they contain nothing, baskets are used as display pieces for decorative purposes, owing to their colour and shape design details.

The colour and shape design details differ widely, depending on the cul-tural setting, the knowledge and experience of the artists, and the market. Various shapes and colours are used. For example, in the Zulu tradition, they may be used in marriage ceremonies as presents. A triangle shape on a

FIGURE 7.3   Painting depicting ceramic pot designs by Andrew Matseba, 2010, Botswana

Source: Photograph by the authors

basket signifies femininity and may express a wish for the young bride to have daughters. Similarly, rectangular shapes represent masculinity. Other shapes and colours would represent health, protection, rain, etc.

Moreover, possessing such an item is believed to have corresponding effects. Many mythology and belief systems are related to the designs and their possible effect. People have a solid sense of relation with the objects, often with faith and religious connotations.

In general, as the artists make the objects for friends, relatives, or the local community, they impart personal touches of craftsmanship, creating bonds between the users and creators. Emphasis is placed on fine quality and mastery of the medium (material and construction). The typical desirable characteristics of the objects are clarity of form and detail, balance complexity of composition, symmetry, and smoothness of finish. The ceramic pot design, as shown in Figure 7.3, is one typical example of the rich artistry of the women of Botswana.

7.6.6   Liberty

For recreation, Afrikans are also outstanding artists. Historical evidence exists in ancient rock paintings in Tsodilo Hills in Botswana. The paintings depict the way of life of the Basarwa (the indigenous people of Botswana): people hunting wild animals, song, and dance. Most paintings have a sense of gaiety, freedom, space, floating, and harmonious relationships between all animals. Historically and artistically, the Basarwa (first inhabitants of Southern Afrika) paintings are some of Botswana's most important cultural works.

7.6.7   Symbolism

The traditional objects, such as the wooden bowl, have smooth, elegant glossy finishes representing cleanliness and good health. A lot of the objects depict some human and other animal forms, interpreting man's life and interdependence with his natural surroundings. Some of the shapes have religious connotations, showing the deep spirituality of the people. Specific people, animals, or the primary forms of spirits are seldom portrayed in the objects. Depending on the cultural context, some of the intricate, symbolic details on some objects would have implied mystic or healing powers.

The sculpture symbolises god, the form is bold with a flowing textured form that drives the viewer to calm face and a bowl. Holistically the figure sculpture looks like a woman god offering food. In Afrikan traditional beliefs, people will go to this god to ask for food during drought periods. They will demonstrate obedience and respect to the god.

7.6.8   Humility

An example of this principle is best explained through the travails of the traditional Afrikan woman in a rural setting. In the Western media, some of the most famous pictures of Afrikan women are those of women carrying babies on their backs whilst typically doing other heavy manual work, such as carrying some load on their heads, with their hands doing some other activities. This would be seen as a sign of admirable strength, but it also has other dimensions. At the centre of the family's daily activities, the woman must draw on her inner strength and rely on her stability in the execution of the daily chores: she must be self-composed, emulate strength and resilience, and always remain humble. In this way, she earns the respect of all, permeating her family and society. Moreover, her family is viewed positively because of her conduct and work.

7.6.9   Resemblance to a Human Being or Animals

To Afrikan artists and art lovers, a carved figure resembling a human has a higher aesthetic value. Artists do not usually portray particular people, actual

FIGURE 7.4   A traditional stone sculpture from Zimbabwe depicting humility, symbolism, composure, and balance

Source: Photograph by the authors

animals, or the primary form of invisible spirits. Instead, they aim to show
ideas about spirituality, reality, or humans and convey them through human
or animal images (Figure 7.4).

7.6.10 Luminosity

The lustrously smooth surface of most Afrikan figural sculptures, often embellished with intricate decorative scaling, indicates beautiful, shining, and healthy skin. Objects with rough surfaces and deformities are aimed to appear ugly and morally flawed.

7.6.11 Self-Composure

This may depict something about the workpiece's creator or owner. The object would be well-ordered and logical and could be related to a composed person: well-mannered and rational; he or she would be controlled, proud, dignified, and elegant (Figure 7.5).

FIGURE 7.5   Hare sculpture depicting self-composure

*Source:*Photograph by the authors

7.6.12   Youthfulness

A youthful appearance connotes vigour, productivity, fertility, and an ability to labour. Illness and deformity are rarely represented because they are signs of evil. Negative aspects of life do not generally feature in Afrikan art.

7.6.13   Clarity of Form and Detail, the Complexity of Composition, Balance Symmetry, and Smoothness of Finish

Afrikan artists value fine quality and mastery of the medium (of material and construction). Because the objects are handcrafted, they can be given detailed personal attention to ensure quality construction and finish (Figure 7.6). The woodcraft skills and techniques are mainly well developed in most societies with a long history of crafts, usually dating back centuries and having been passed from one generation to the other.

FIGURE 7.6   Coasters container

Source: Photograph by the authors

7.7 Afrikan Cultural Symbols and Artefacts

Gonzalez (2018) describes a cultural symbol as a physical manifestation that signifies the ideology of a particular culture or that merely has meaning within a culture. Aziza (2001, p. 31) refers to culture "as the totality of the pattern of behaviour of a particular group of people". It includes everything that makes them distinct from any other group of people. For instance, their greeting habits, dressing, social norms and taboos, food, songs and dance patterns, rites of passage from birth through marriage to death, traditional occupations, and religious and philosophical beliefs.

After studying the meaning of artefacts from dictionaries through their evolution from early 1800, Friedman (2007) proposes a broadly inclusive meaning. He quotes Simon (1982), saying, "an artefact is anything that we can design in the huge sense of the word design, defined as '(devising) courses of action aimed at changing existing situations into preferred ones' (Simon, 1982, p. 129)".

Afrikan cultural symbols and artefacts are representations and objects which pertain to the Afrikan way of being. Ezedike (2009, p. 455) defined Afrikan culture as: "the sum of shared attitudinal inclinations and capabilities, art, beliefs, moral codes, and practices that characterise Afrikans". It can be conceived as a continuous, cumulative reservoir accommodating material and non-material elements that are socially communicated from one generation to another. Afrikan culture, therefore, refers to a whole lot of Afrikan heritage.

A representative example is the Adinkra symbols and their meanings (Table 7.1). Originating from West Afrika, e.g., Ghana and the Ivory Coast, the Adinkra are visual symbols used by the people of that region over centuries. They were typically drawn or printed on textiles and initially worn by high-ranking clan people and royalty at ceremonial events and rituals. They have always had particular importance and significance, whether religious, mythical, philosophical, or otherwise. The present-day use of Adinkra symbols is global, with variegated use in interior designs, furniture, different graphic design works, ceramic works, clothing, textiles, applied arts such as sculpture, and many more. These symbols are embedded with meaning that relates to the peculiarities of the people's culture. Table 7.1 shows a sample of the Adinkra symbols and their meanings.

7.7.1 What Are Symbols?

"A symbol is a visual image or sign representing an idea – a deeper indicator of universal truth" (Bruce-Mitford, 2008, p. 6). Womack (2005) further describes symbols as a means of complex communication that often can have various stages of meaning. Symbols are representations of an entity; a human being creates to use and interpreted by fellow human beings. Therefore, they

TABLE 7.1 Typical Adinkra symbols and their meanings

Symbol	Name	Meaning
	Sankofa	Wisdom of learning from the past
	Osram ne Nsoromma	Faithfulness, fondness, harmony, benevolence, and love
	Adinkrahene	Authority, leadership, and charisma
	Akoma	Love, goodwill, patience, faithfulness, and fondness
	Nkonsonkonso	Human inter-relations and bonds
	Gye Nyame	Omnipotence of God
	Nyame Dua	God's presence and protection
	Dwennimmen	Strength (in body, mind, and soul)
	Nteasee	Understanding and cooperation
	Aya	Endurance, independence, perseverance, and resourcefulness
	Nsoromma	Faith and the belief in patronage and dependency on a supreme being
	Adwo	Peace, tranquillity, and quiet

are created to convey a specific significance, meaning, or value to the person beholding them. Together with signs, the study of the significance of signs is essential in the cognisance of human–object interaction and semiotics.

7.7.2 Afrikan Artefacts

The furniture pieces in Figures 7.7 and 7.8 were designed and made in Afrika, depicting modern product form through semiotic understanding and appreciation, combining modern and traditional materials, and using the elements and principles of the Afrikan aesthetic, such as craftsmanship, symbolism, self-composure, and togetherness. The furniture pieces use rich colourings from local soils and patterns taken from traditional textiles.

FIGURE 7.7 Fofa sofa design

Source: Molokwane and Steyn; Photograph by the authors

FIGURE 7.8 Iketlo sofa design

Source: Molokwane and Steyn; Photograph by the authors

7.8 Application of Afrikan Elements and Principles in Design

Afrikan design is a rich and complex field that incorporates a wide range of elements and principles to create visually striking designs, culturally meaningful and artistically compelling (Figures 7.9 and 7.10). The dress fabric and placemat design were inspired by a traditional basket using specific design principles. The patterns in the basket symbolise a running ostrich – an inspiration from fauna. When ostriches chase each other, they run in a zig-zag movement, depicted in Figure 7.9 in an abstract form. Traditional designs have used principles such as movement, represented by flowing lines and shapes suggesting motion and energy. The text design in Figure 7.10 also represents movement. The patterns in the dress symbolise the unity of bringing different elements together to celebrate shared values and traditions. The repetition of patterns creates emphasis to draw attention to specific elements or motifs, thus, creating a focal point that draws the observer's eye. The dress incorporates balance and symmetry elements with a visually balanced and harmonious design that uses scale and proportion to create visual interest and balance.

The design creates a sense of balance and coherence through the design elements of colour, form, and texture. The colours used in baskets are organic and sustainable obtained from cooking roots and/or leaves of different species of trees. This process produces earth colours, as depicted in the colours in Figure 7.9. The earth colours in Afrikan design are profoundly symbolic and reflect the close connection between human life and the natural world. They represent the foundational elements of life, the power of nature, and the connection between the living and the spirits of ancestors. For example, in many Afrikan cultures, brown represents the earth, strength, and stability. It is often used in Afrikan design to symbolise the foundation of life, the ground that sustains human life, and the stability of family and community. The colour red symbolises life, vitality, energy, and the power of ancestors. It is also associated with power, passion, and courage. It is associated with happiness, joy, and prosperity. In Afrikan design, yellow is often used to symbolise the sun, the source of light and warmth; warmth; fertility; and the life-giving force that sustains people. The texture in Afrikan design is an essential part of symbolism that conveys specific meaning. Patterns and colours communicate wisdom, wealth, prestige, protection, fertility, or strength. For example, Figure 7.9 shows a prestigious dress, and the placemat acts as a table protection device (Figure 7.10). Designers manipulate textures to communicate ideas, emotions, and values, making texture an essential part of Afrikan design.

FIGURE 7.9   Fabric design inspired by basket weaving

Source: Photograph by Keiphe Setlhatlhanyo

FIGURE 7.10 Table placemat inspired by basket weaving

Source: Photograph by Keiphe Setlhatlhanyo

7.9 Harmonisation of Interpretations: Elements and Principles of Design

The different interpretation of elements and principles of design is culturally influenced. This influence is also in the aesthetics of products, where elements and principles are used primarily to drive the product appearance agenda for its wide adoption by homeowners and architects (Awuku et al., 2021; Brako & Gilbert, 2022; Vinchu et al., 2017). The interpretation of elements is also contextual, giving meaning to how they are being used to communicate a composition, product, space, or environment. Curnow (2018) discusses the elements and principles of design with an Afrocentric take on the Eurocentric elements. Therefore, the understanding of Afrikan aesthetics has a Westernised reference due to the hybridisation of cultures. This hybridisation of cultures is critical in exchanging knowledge and skills, sources of inspiration for design work and cross-culture lessons. In the introduction section of this chapter, universally and widely used elements and principles of design were described from the perspective of Afrikan aesthetics. The development of this chapter brings another dimension to the interpretation of elements and principles of design in Afrikan aesthetics, further extending their meaning and their extent of use. What is shared between the cultures as elements and principles are shown in Tables 7.2 and 7.3.

What are the fundamentals of all design, Afrikan, European, Asian, or American? What is unique in the Afrikan context that can be shared with the rest of the world? The fundamental elements and principles are the same across cultures. The interpretation varies, as demonstrated in Section 7.3, where various interpretations and uses are exemplified with products. Afrikan symbolism discussed in Section 7.4 can be transferred to another context to enrich design experiences with lessons from Afrikan cultures.

TABLE 7.2 Elements of design

Widely Used Element	Western/Eurocentric	Afrikan
		Symbolism
Line	Line	
Shape	Shape	
Form	Form	Form
Size	Size	
Space	Space	
Colour	Colour	Colour
Texture	Texture	Texture

TABLE 7.3 Principles of design

Widely Used Principle	Western/Eurocentric	Afrikan
Unity	Unity	Unity
Harmony	Harmony	Harmony
Rhythm	Rhythm	Rhythm
Balance	Balance	Balance
Proportion	Proportion	Proportion
Emphasis/dominance	Emphasis/dominance	Emphasis
Transition/Movement	Transition/Movement	Movement
Variety	Variety	
Contrast	Contrast	

7.10   Conclusion

This chapter postulates that although we have been interpreting design elements from a Eurocentric reference, Afrikan design is informed by diverse cultures and lifestyles. Product design and development have existed in the continent in informal contexts. The people have always painted, e.g., San rock painting, and made tools and implements for their activities, albeit they need industrial infrastructures and production systems. The Afrikan design and aesthetics principles have been related to the arts and crafts developments, religious beliefs, and cultural references indigenous to Afrika. Significant Western designs, such as the art deco style, have been influenced for a long time by the materials, forms, and techniques of Afrikan arts and crafts. Contemporary arts and crafts products and processes globally use the Afrikan influences in endless configurations and significant proportions.

The ethos of Ubuntu as a uniquely Afrikan ideal which applies to global spaces must be about recognising and accepting that Afrika has a rich base of inspiration and resources, which, when well intimated with and appropriately used, offer limitless opportunities for creating the best for humanity. There are several aspects to using Ubuntu-based design elements and principles for designing better products and services. It is about a renewed perspective of a co-created design centred on people. It is about using the cultural contexts and local nuances, leveraging the commonalities and differences where relevant, local resources and practices, and blending sensitively with modern technologies to create effective integrated working processes and systems. This can lead to resounding results, not only for the local Afrikan context but with the right impact on the global stage.

The challenge for Afrikan designers is to explore these influences for the best advancement and contribution of the design discourse for Afrika and the world. The creatives would explore, amongst others, the indigenous materials, handcrafts, and artisan cultures, and cultural-oriented design approaches informed by myths, folklore, music, dance, etc. The collective work by Afrikan academics, artisans, consultants, promoters, sponsors, etc. would make Afrikan design, especially industrial design, and other disciplines asserting themselves well on the global stage.

Reference List

Abusabib, M. A. (1995). *African Art: An Aesthetic Inquiry*. Uppsala: Academiae Ubsaliensis.

Adams, E. (2013). The elements and principles of design: A baseline study. *International Journal of Art & Design Education*, 32(2), pp. 157–175.

Awuku, S. A., Bennadji, A., Muhammad-Sukki, F., & Sellami, N. (2021). Myth or gold? The power of aesthetics in the adoption of building integrated photovoltaics (BIPVs). *Energy Nexus*, 4, 100021.

Aziza, R. C. (2001). The relationship between language use and survival of culture: The case of Umobo youth. *Nigerian Language Studies*, 4, pp. 31–41.

Brako, D. K., & Gilbert, S. J. (2022). Elements and principles of design in scenic design in Ghana's most beautiful reality TV show: An aesthetic evaluation. *Journal of African History, Culture and Arts*, 2(2), pp. 83–93.

Brooks, L. (1971). *African Achievements: Leaders, Civilisations, and Cultures of Ancient Africa*. Jacksonville: De Gustibus Press.

Bruce-Mitford, M. (2008). *Signs & Symbols: An Illustrated Guide to Their Origins and Meanings*. London: Dorling Kindersley Limited.

Curnow, K. (2018). *The Bright Continent: African Art History*. Cleveland: MSL Academic Endeavors.

Edel Jr., D. H. (1967). *Introduction to Creative Design*. Englewood Cliffs, NJ: Prentice Hall.

Evans, P., & Thomas, M. (2012). *Exploring the Elements of Design*, 3rd ed. Delmar, NY: Cengage Learning.

Ezedike, E. O. (2009). *African Culture and the African Personality. From Footmarks to Landmarks in African Philosophy*. Somolu: Obaroh and Ogbinaka Publishers.

Friedman, K. (2007). Behavioural artefacts: What is an artefact? Or who does it? *Artefact*, 1(1), pp. 6–10.

Gatto, J. A., Porter, A. W., & Selleck, J. (2011). *Exploring Visual Design: The Elements and Principles*, 4th ed. Worcester, MA: Davis Publications.

Gonzalez, K. (2018). *Cultural Symbol: Definition & Examples*. https://study.com/academy/lesson/cultural-symbol-definition-examples.html.

McClurg-Genevese, J. D. (2005). The principles of design. *Digital Web Magazine*, p. 13.

Molokwane, S. (2007). *The African Aesthetic as It Informs the Product Form*. Universidad Iberoamericana. Ciudad de México. www.dis.uia.mx/conference/2007/ponencias_i.php.

Okpewho, I. (1977). Principles of traditional African art. *The Journal of Aesthetics and Art Criticism*, pp. 301–313.

Simon, H. (1982). *The Sciences of the Artificial*. Cambridge, MA: MIT Press.

Soğukkuyu, B. (2021). *Analysis of Poster Designs of Turkish T.V. Series on Ottoman History: Resurrection Ertugrul and Magnificent Century Examples*. Hershey: IGI Global.

Van Sertima, I. (1983). *Blacks in Science: Ancient and Modern*. New Jersey: Transaction Publishers.

Vinchu, G. N., Jirge, N., & Deshpande, A. (2017). Application of aesthetics in architecture and design. *International Journal of Engineering Research and Technology*, 10(1), pp. 183–186.

Vogel, S. M. (1986). *African Aesthetics*. New York: Center for African Art.

Womack, M. (2005). *Symbols and Meaning: A Concise Introduction*. Oxford: Altamira Press.

Yeo, A., and Cao, F. (2021). A creative research process for a modern African graphic design identity; The case of Ivory coast. *Art and Design Review*, 9, pp. 210–231.

8

DESIGN SKETCHING

What Works Best From an Afrikan Perspective?

Yaone Rapitsenyane and Matthews Ollyn

Introduction

Design sketching in higher education design programmes across Afrikan universities has adopted conventional teaching and learning approaches. Design sketching is a critical communication and creative tool in the design process, mainly when translating design research insights into ideas. As a communication tool, sketching can be regarded as an intrinsic design language between individuals during the design process and subsequent uses of the generated sketches. As it is being done today, design sketching is a representation of mental imagery in drawings detailed enough to guide the design process. Afrikans directly translated these mental images into products without translating them into sketches using pencil and paper.

The Ubuntu principle of perseverance, which speaks to strength, commitment, and cohesion (Mangaroo-Pillay & Coetzee, 2021), has provided courage for many Afrikans from disadvantaged backgrounds to get through hardships of life, including earning an education. With this background and approach to design sketching, the widely acknowledged challenge of sketching inhibition (Thurlow & Ford, 2018; Rapitsenyane et al., 2019; Thurlow et al., 2019) is being addressed through human-centred design. Through human-centred design, the design education environment becomes sensitive to the learner and context and adopts empathy and value co-creation to sketching to relate in-class learning to the context of shared social interactions (Chmela-Jones, 2017). Human-centred design can further be applied to the problems investigated during the early stages of design education. Therefore, more time can be spent on finding the solution rather than understanding the problem.

DOI: 10.4324/9781003270249-12

This chapter aims to detail a sketching approach which can be adopted from an Afrikan point of view as practised at an Afrikan university. The chapter provides an understanding of the cognitive aspects of sketching, appropriate cognitive behaviours being developed by sketching and the general approach to sketching, which reduces the difficulty of sketching. Empirical evidence from how to sketch in design is taught at the University of Botswana, Botswana. The Department of Industrial Design and Technology at the University of Botswana has revised its two undergraduate programmes – Bachelor of Industrial Design and Bachelor and Design and Technology Education – to meet the needs of various stakeholders. Among the objectives of the revision was to ensure fitness for purpose and suitability for the market. Design-related practice in the country is still crafts-based, and professional design practice still needs to be developed. Sketching as one of the critical design skills, was specifically accommodated and strategically prioritised in the revisions of the programmes.

Sketching Cognition

Sketching is still a powerful tool for designers to develop design ideas, generate concepts, externalise and visualise problems, facilitate problem-solving and creative effort, and revise and refine ideas (Coley et al., 2007). Tytler et al. (2019) describe sketching as a constructive reasoning and learning process. In design, it is one of the tools supporting the problem-solving process. Dulic and Krkljes (2018) view design sketching as a process of creative mental synthesis and that plays a central role in product development. This is because sketching combines several cognitive processes, such as attention, perception, learning, remembering, speaking, problem-solving, reasoning, and thinking at different application levels. Such memory operations as registration, storage, and retrieval become critical in transforming visual sensory inputs into recognisable formats, holding information in memory, and extracting it during problem-solving tasks (Rapitsenyane et al., 2018). Sketching holds fascinating cognitive information about designers when closely examining the sketching behaviours (Coley et al., 2007; Suwa et al., 1998). Sketching has been found to help designers enhance the variety and originality of their exploration, and the reasons for this observation still need to be understood (Brun et al., 2016). Observing how designers cognitively interact with their sketches can inform people's essential behaviours to cultivate through sketching. Several researchers have reported these behaviours – Musta'amal et al. (2009) report them as combining, restructuring, lateral transformation, vertical transformation, part-by-part drawing, non-part-by-part drawing, and reflective and experimental behaviour.

The latest debate on trends in sketching emphasises the quality of the information generated instead of the quantity of the information (Elsen et al.,

2012). This debate is tied to the timing of use and the use of sketching in the design process. This contradicts an earlier debate emphasising quantity and building correlations between the quantity of brainstormed ideas and design outcomes (Yang, 2009). In the perspective advanced by Yang (2009), the effect depends on the tool used to generate ideas. However, both Elsen et al. (2012) and Yang (2009) concur that the timing of the use of sketching in the design process is crucial to its success as an idea-generation tool. The advantages of sketching in this way are still more inclined towards manual sketching (Visser, 2006; Suwa et al., 1998; Bilda & Gero, 2005; Cross, 2000; Leclercq, 2004; Détienne et al, 2004). Cognitively, manual sketching has also been advantageous over digital sketching in several ways. Cross (2006), as quoted in Hua et al. (2018), classifies design sketches into two groups: a working sketch and a non-working sketch. A working sketch refers to a group of sketches produced by designers during the product design process, while the latter is produced in their spare time outside the design process. This difference defines how much the sketches can be used in the creative process. Non-working sketches can capture and communicate ideas which can only be understood by the designer.

Despite the development of digital sketching tools, manual sketching is still widely used because of the advantages it possesses to date. Hua et al. (2018) explain that before a digital design is applied during the product design process, design outcomes are typically presented as sketches to communicate with other interested parties. Early design activities are, therefore, still manual sketching driven, with the incremental introduction of digital tools as more detailed and refined design information is processed in the downstream design and prototyping activities. Sketches are increasingly being adopted in other fields to help facilitate better communication. Tytler et al. (2019) have positioned sketching as a potential essential resource for reasoning and learning in science classrooms. Schwartz (1995), as quoted in Tytler et al. (2019), further notes the benefits of collaborative drawing, in which the specificity and explicitness of drawings provide opportunities for peers to exchange and clarify ideas.

Art and Design Sketching: Reducing Inhibition

Principal to reducing inhibitions is the philosophical approach of Ubuntu and the concept of mashed-up aesthetics as techniques to learn from Afrikan contexts when teaching Afrikan student designers. Sketching inhibitions can be reduced using art-based exercises early in design-sketching classes (Rapitsenyane et al., 2022). This approach allows students to gain confidence in sketching freely with few technical ground rules. There are a lot of easy-to-draw Afrikan artworks, symbols, and crafted products, of which, when used as warm-up exercises to design sketching, can improve students' confidence

(Rapitsenyane et al., 2019). Traditionally Afrikans, like other cultures, have always used symbols for communication and interaction where words could not (Umeogu, 2012), reaffirming that a picture says a thousand words. Today, drawings/sketches have taken on the role of the symbol with more roles, including optimisation and planning of products. Afrikan-crafted products often reflect people's identity more than forging relationships with the markets (Plug & Collins, 2022). Such an approach forms sentimental and identity value across generations and can be quickly retrieved in Afrikan mental models of creative and innovative products, hence effective as examples in a product design class.

Visual Communication in Design

Dulic and Krkljes (2018) describe sketches and other types of drawings as essential tools the designer uses to clarify ideas and ensure the design team possesses the same mental model. Furthermore, they observed that visual communication at the beginning of the design process is mainly through the unstructured use of sketches. Sketching is commonly regarded as a critical aspect of creative thinking in design (Lawson, 2002) and a significant part of design pedagogies (Oxman, 2008). Design students are taught to think through sketching, which can help them to externalise concepts, communicate ideas, and solve complex problems (Hua et al., 2018). Furthermore, the findings of the experiment by Hua et al. (2018) suggest that sketching skills are linked to design creativity. Less abstract and more realistic representations describe the later stages of the design process. Dulic et al. (2018) further assert that the primary purpose of a drawing is to describe the design so that those for whom it is intended can implement it. Thus, the audience and, to some extent, the design process stage influence the visual communication (drawings) produced.

On the other hand, digital design is described as a process in which design decisions are made 'on screen' rather than with sketch paper through all stages of the design process (Marx, 2000). Although they seem different and distinct, they have a common goal: to make representations for conceiving and communicating in the conceptual design stage. Tovey et al. (2003) argue that manual sketching is a thinking tool, while the digital design is a communication tool. The same view is upheld by Self (2013), who asserts that both tools are relevant and crucial and demand that the designer tap into their experience and use them relatively and appropriately. The timely integrated use of the two often results in a highly resolved design that can be rendered at photorealistic quality, often perceived as authentic by the viewer (Ranscombe & Bissett-Johnson, 2017). Traditional representations are far more fluid and appropriate than digital media for the initial and fast development of ideas and the stimulation of the imagination. Traditional representations also allow designers to be closer to their context as a quick reference for inspiration.

Design Sketching – the University of Botswana Approach

In the structure of the design programmes at the University of Botswana, sketching and related skills come at the foundation levels of years 1 and 2. As a build-up to learning how to sketch, students are exposed to creative thinking techniques and learning by doing at year 1 in a course called Design Fundamentals. At the same time, students learn about elements of design such as space, line, colour, shape, motion, texture, pattern, and value, and principles such as balance, rhythm, proportion, dominance, scale, and unity (Ziff, 2020), as an organised way to talk about visual literacy. This organisation is vertically integrated with courses focusing on manual sketching late in the first year and digital sketching techniques and styling early in their second year of study as the student's confidence grows. During the manual sketching course, a less rigid approach is used to get students excited about sketching and reduce sketching inhibition.

During the introduction of the manual sketching course, the course aims and objectives and the course tutorials are launched to give the students the course expectations. In the first two weeks of the course, art-based warm-up activities found to be effective in reducing sketching inhibition (Hu et al., 2015) are used to build students' confidence. This approach is repeated when students are introduced to digital sketching techniques and styling through another art-based exercise called mashed-up aesthetics. Observational drawing is used before manual sketching and teaching principles for approaching a design sketch. The artful thinking model is used for this part of the course to develop fundamental skills of observation, identification, and early application of elements and principles of design. This approach is used through structured tutorials, fully described, and discussed as the students review the drawings. Artful thinking was developed by Project Zero at the Harvard University Graduate School of Education to 'help students learn how to think by looking at and exploring works of art' (Barahal, 2008, p. 299).

Art-Based Warm Up Exercises

A quick recap on elements and principles of design and design fundamentals, which the students would have done in the previous semester, is done to provide focus and perspective for the course and the structured tutorials. At the end of the class, where students had a first attempt (Figure 8.1), a class discussion is held to allow the students to show their work and discuss how they progressed through it, including difficulties in identifying and interpreting elements and principles of design in physical objects and pictures brought to class. This active participatory engagement of students also allows reciprocity between two vertically integrated courses such that students can start transferring knowledge and skills from one course to the other. The objects of observation for the warm exercises are pictures of Afrikan crafts, such as Setswana traditional chairs (Figure 8.1), stools, basketry, and pottery products, which would later require various complex design sketching techniques to draw. These products are used since almost all the students in the class would usually be of Afrikan origin and relate to them as known visual stimulation cues. After an effective review of this approach with enhanced students' confidence in holding and moving the pen, a structured design sketching approach is introduced in the same format as structured tutorials, ending with a semester-long project. This is repeated in the next-level course for digital sketching and product styling, where mashed-up aesthetics is used as students continue to build on pre-requisite knowledge and skills.

Structured Design Sketching Tutorials

Structured design tutorials introduce the technical side of design sketching once the students build their confidence. The tutorial objects are also focused on contemporary consumer products appropriate to impart a particular sketching skill.

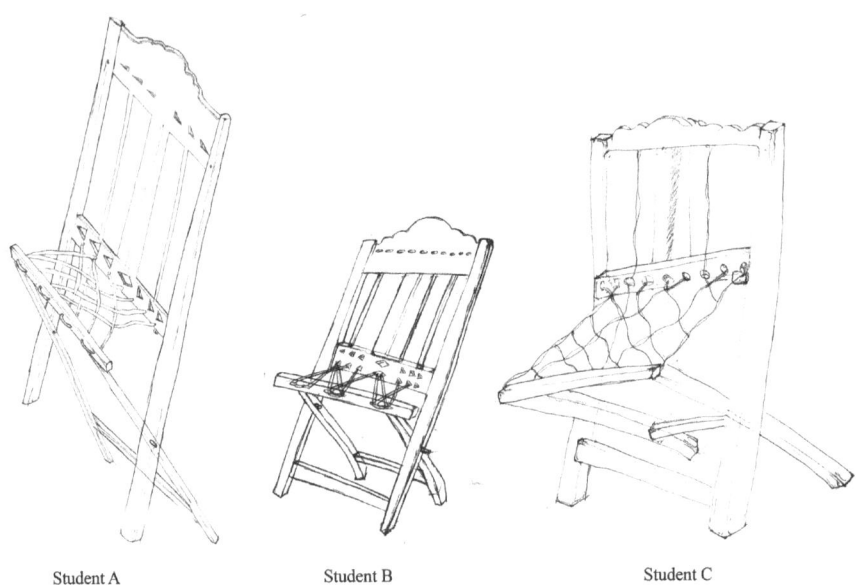

Student A   Student B   Student C

FIGURE 8.1 Students' first attempt at sketching through observation a Setswana traditional chair (Year 1, 2022/2023 students)

Design Sketching Exercises

From the evaluation of art-based exercises, lines are the first to pick on, and students are taught how to join two points with straight lines (Figure 8.2) and sketch curves with points on a smooth trajectory (Figure 8.3). During sketching, these exercises train hand-eye coordination, concentration, speed, and accuracy.

Crating Tutorials

Objects in real life appear to be varnishing or reducing in size as they get away from our eyes. Edges and surfaces closer to our eyes appear broader and more prominent, and those farther from our view appear smaller. This is called perspective view in sketching. Products are represented through perspective sketching. The principle is brought to life by analysing different products, big and small. Crating is the first step from side-view sketching to sketches in perspective. The product is first explored and understood through elevation sketches, which help determine the best approach for the sketch. A perspective box (rectangular cube) is usually the starting point after introducing the concept of perspective sketching. Products with basic geometries, such as rectangles, squares, and triangles, are then crated. Emphasis is on showing construction details to demonstrate the principles of sketching in perspective. A sketch deploying the crating method is shown in Figure 8.4.

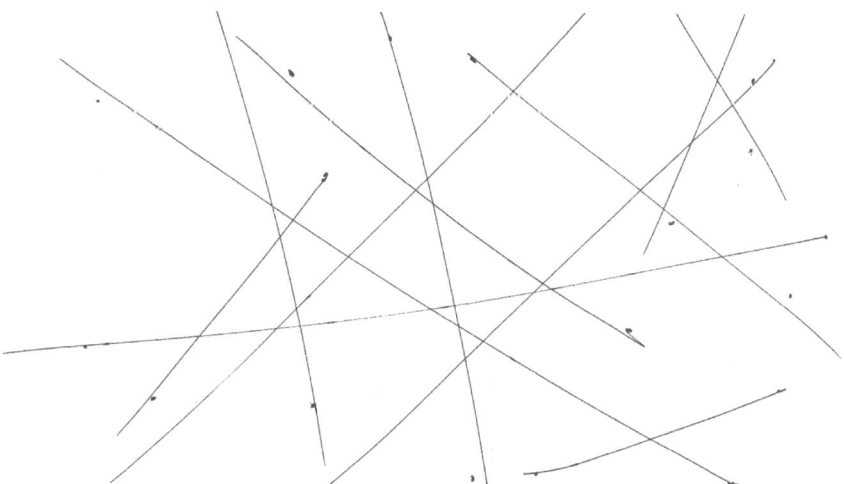

FIGURE 8.2   Joining two points with a straight line

Source: Authors' in-class demonstration

FIGURE 8.3 Joining points lying on a smooth curve

Source: Authors' in-class demonstration

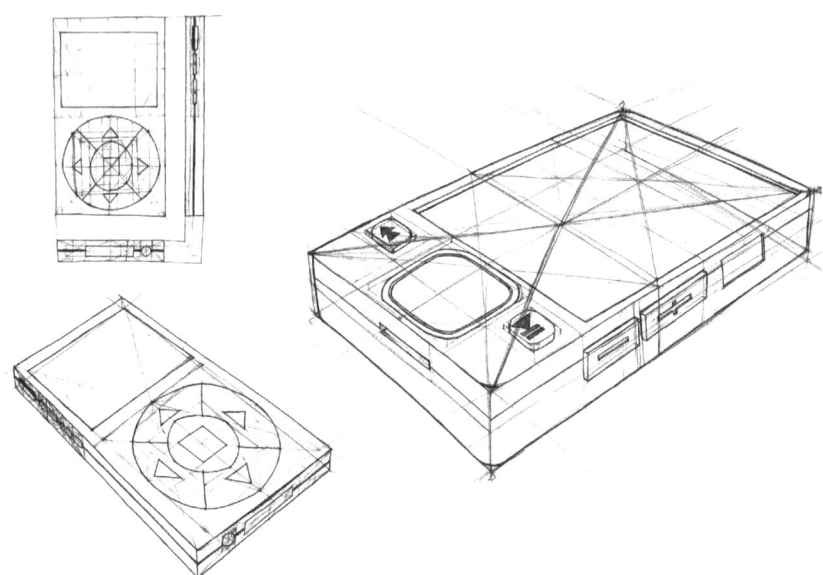

FIGURE 8.4 Crating tutorial submission (Year 1, 2021/2022 student's work)

Approaching Ellipses

After students understand crating, they are introduced to ellipses and elliptical forms. The approach is based on the analysis that ellipses have axes (the
major and minor axis) and that these axes can be used to determine the direction of the ellipse on a sketching plane and place elliptical forms in other
forms. The principle of using axes to construct ellipses is that the major axis
and the minor axis cross always intersect at 90 degrees and that the revolv-
ing axis of the ellipse always lies along the minor axis. With these principles,
elliptical forms can be sketched on the right plane to sit well with the other
forms communicated in the design (Figure 8.5).

Sketching Complex (Spherical) Forms

Complex spherical forms which cannot be crated can also be approached
by determining the plane on which they sit so that they are sketched at
the right location and facing the right direction in the right proportion
and scale in the design (Figure 8.6). Sketching complex forms deploys
sketching planes the same way as in parametric modelling environments.
This allows designers to express organic forms which otherwise cannot be
crated.

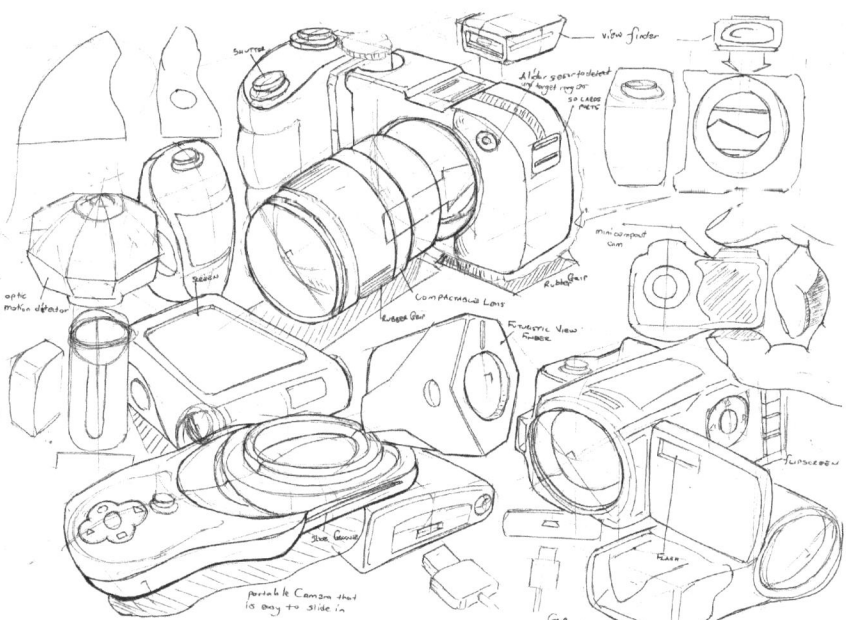

FIGURE 8.5 Sketching ellipses (Year 1, 2021/2022)

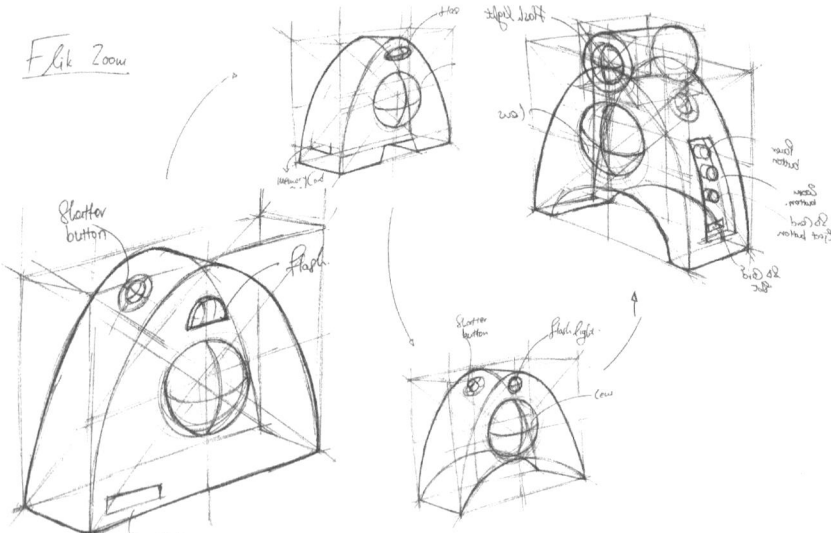

FIGURE 8.6 Complex spherical forms combined with ellipses and crating (Year 1, 2021/2022)

Semester-Long Project

A culmination of the exercises and tutorials in the sketching class is a semester-long sketching project to combine all the skills acquired during the course. The project is also structured to allow all students to cover critical sketching approaches and presentation techniques and assess the execution of similar or related skills with less subjectivity. Design fundamentals are applied in the creative part of the project to ensure the outcomes are different. Each student is expected to produce a mood board as a collection of their sources of inspiration to stimulate the creative part of the project. The assessment criteria then reflect all aspects covered during the semester to guide the students towards a holistic understanding of design sketching (Table 8.1).

Structured Hybrid Sketching Projects

In order to maintain the sketching momentum and help students relate learning across courses and levels of their learning programmes, an integrated approach to teaching and learning is adopted. A curriculum is integrated if it connects different areas of study by identifying and spelling concepts that unify subject matter lines, both vertically and horizontally (Magoma, 2016). This approach allows students to relate more closely to the content and become more actively engaged in their learning. The articulation of sketching courses at the Department of Industrial Design and Technology permits

TABLE 8.1 Design sketching assessment criteria

Component	Description	Possible Marks	Actual Marks
Perspective	The perspective is perfect; the sketches clearly show that they have been drawn on 1-point, 2-point, or 3-point perspectives as per the designer's preference and the task's need. Views are chosen to show the sketches in perspective to their best advantage. Sections and contours have been appropriately and adequately used to express form.		
Quality of lines	The line work consists of different types of lines to define borders and details within the line work. Outlines and thick and thin lines have been appropriately used to make the sketch prominent/'stand out'. Lines for thumbnails are also continuous, NOT broken short, and there are no untidy pen marks. Different pen colours have also been used to define and emphasise form.		
Page anatomy	The page has been creatively used. The page layout has been planned, and sketches have not been randomly placed on the page, cluttered on one side, or scattered in a way that shows poor use of negative space. Where text has been used, it is neat and parallel to the paper's horizontal edge. Sketches vary in size and views. There is a good combination of side views, big and small perspective sketches, construction, and detailed sketches. Skills such as grouping and other enhancements have also been applied to demonstrate various design elements.		

(Continued)

TABLE 8.2 (Continued)

Component	Description	Possible Marks	Actual Marks
Approach	The right approach has been used for varying complexity of the product (symmetrical, basic geometry, and complex organic forms): Side views Primitives Crating Contours Sections		
Enhancements rendering	Shadows have been used to give an impression of light falling onto the object from a particular direction; groupings show components of an idea; context gives some sense of scale, etc. There is a feel of light falling onto the object, clearly visible by the effects of the renderings. The poor renderings do not degrade the quality of the sketch.		
Logic	The work is logical and shows improvement from the initial to the final sketches.		

vertical curriculum integration. From the manual sketching course, the students use their outcomes as input learning materials for the digital sketching course, with their sketches now as bitmap images. This hybrid sketching approach demonstrates the valuable interplay between manual sketches and digital sketching techniques in the design process (Rapitsenyane et al., 2022). Quick presentation of ideas to users and potential customers can be produced from these hybrid sketches within a short period, instead of a CAD model, which takes longer to produce. The bitmap image is processed to a high photorealistic quality in Adobe Photoshop (Figure 8.7).

Structured hybrid sketching projects are extended in learning product styling. A theme is usually provided for the styling project. At the same time, activities and content are delivered to provide an understanding of styling and how to go about styling products. The sketching techniques and skills are transferred to these projects and assessed so that students emphasise constant improvement. This is usually done in their fourth semester, and the quality of the manual and digital sketches is close to that of professional designers (Figures 8.8 and 8.9).

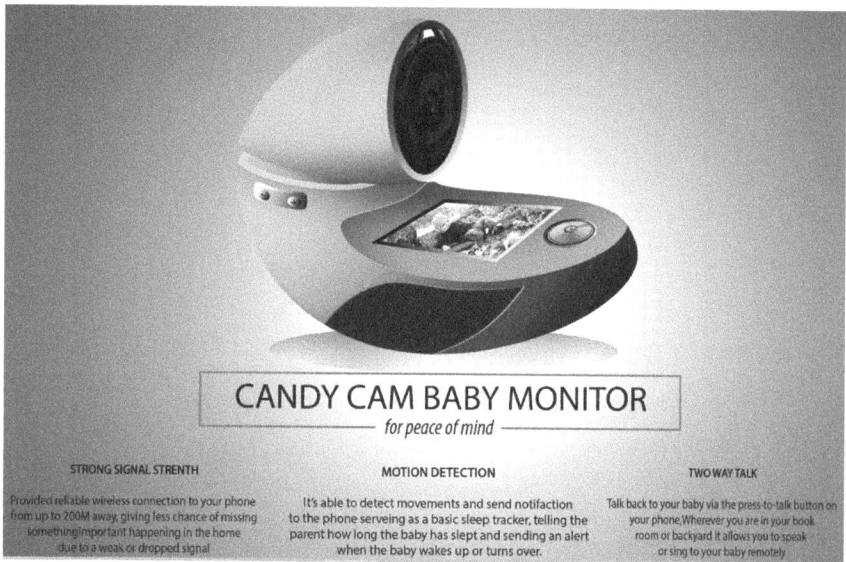

FIGURE 8.7 Digital rendering of a bitmap image in Adobe Photoshop (year 2 student in 2018/2019)

FIGURE 8.8 Styling a Lexus water bottle (manual sketches by a year 2 student in 2018/2019)

FIGURE 8.9 Rendering a Lexus water bottle (digital sketch by a year 2 student in 2018/2019)

The approach to sketching products of varying sizes allows the transfer of knowledge and skills from generic sketching courses to styling courses. Products as big as vehicles are also styled after learning approaches to how to sketch cars. Even though transportation design is not a focus in the industrial design programme, the range of products and experiments possible with sketching are usually an opportunity to challenge students' ability to comprehend abstract concepts when applied to bigger products (Figure 8.10).

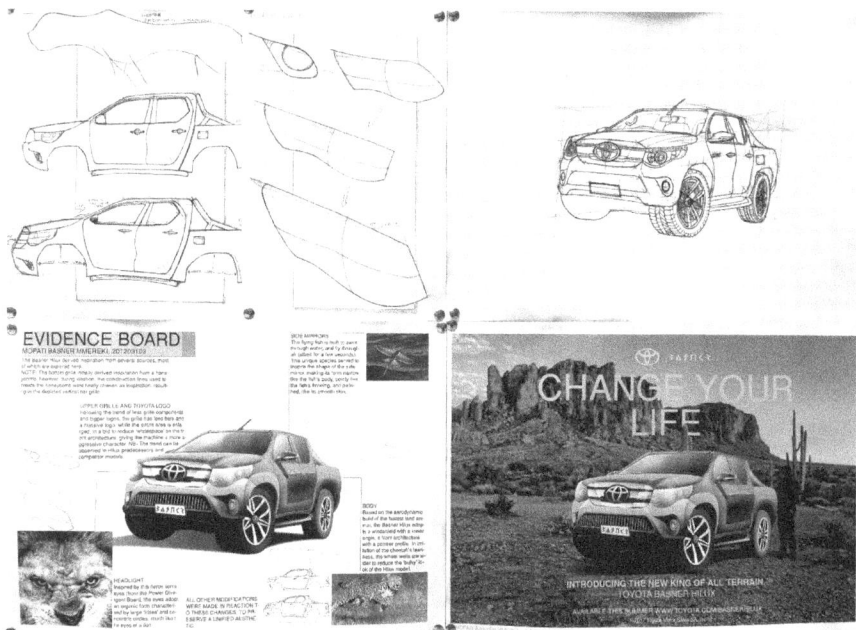

FIGURE 8.10 Vehicle styling project (year 3 student in 2015/2016)

Mashed-Up Aesthetics

Mashed-up aesthetics is one of the techniques used to kick-start the ideation process. Students start the creative process by merging unrelated digital 2D images to create a form of the product to style. Following the demonstrations and exercises on mash-up aesthetics, students can create inspirations to drive their form development. It is with this mashed-up image that idea generation starts. After several sketching and foam modelling iterations, the final designs are scanned in an A3 scanner and rendered digitally in Adobe Photoshop. Sample students' work in this process is shown in Figure 8.11. These activities are driven by a lot of sketching, even though the focus is on creative thinking in product design. The ideas are usually diverse as students sketch even from the same source of inspiration. Most students can define the overall form of the products and positioning of key features using sectioning, contour lines, and detailing in their sketches. The learners display an insightful aptitude to digitally enhance their sketches, as most of them can present work with a good balance of rendering considerations for the digital manipulation of bitmap images. However, in some instances, many do not display good intuition in using Adobe Photoshop tools to develop their renderings. This is because the students would still

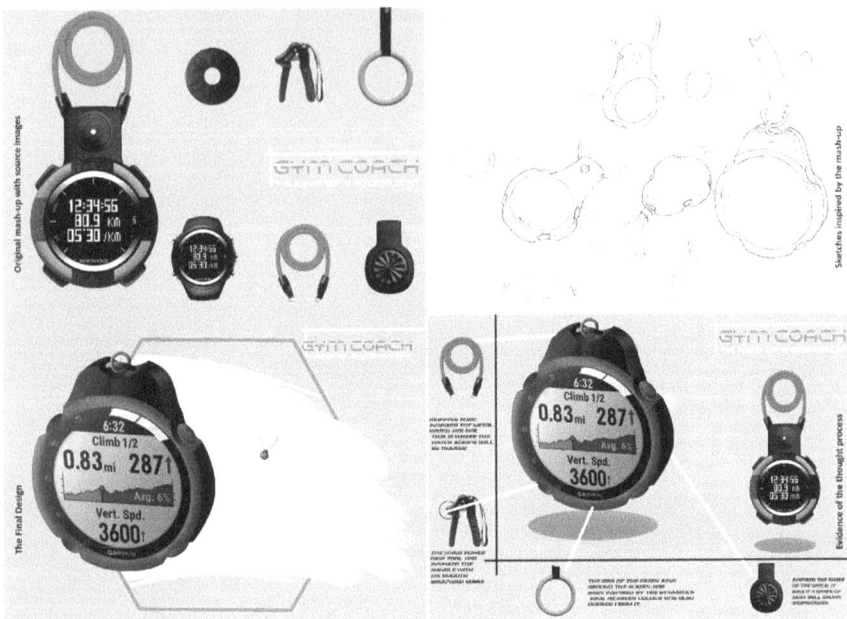

FIGURE 8.11 Sample mashed-up work (Year 2 student in 2019/2020)

practise using digital design communication tools. Most of the renderings would appear crude and unpolished, which suggests that they had not yet reached a good level of proficiency with using the tools in the digital platform. However, digital rendering is advantageous in some poorly developed manual sketches. This gives students a second chance to correct some aspects of their sketches, such as the accuracy of ellipses, accuracy of the perspective and level of detail in defining product features. The result is good-quality renderings which in some cases are very much divorced from the manual sketches.

A Design Sketching Philosophy

The approach to this chapter was to articulate the need for a sketching philosophy in a design programme or organisation. Philosophical considerations include tacit contextual knowledge, which facilitates learning technical design sketching. Art and craft are prominent in most Afrikan markets, and most Afrikan students would have come across one or many of these products as they grew up, including their production processes. This is a direct stimulus to understanding how this product could be perceived, interpreted, and presented in a 2D media, hence art-based exercises (Rapitsenyane et al., 2022). Interacting with Afrikan crafts at an artistic level is a building block towards the right design sketching mental model. This is especially true for design, as early visual communication is usually made through something other than structured sketches (Dulic & Krkljes, 2018). Thinking with sketches corroborates the artful thinking model, where learners go through routines that help them focus on the specific aspects of the object they observe at the start of the sketching course.

Elements and principles of design are critical in successfully implementing a sketching philosophy. Sketching, as some form of visual communication, depends on the interpretation of elements and principles of design. A structured way to assist learners in understanding elements and principles is critical to the success of any sketching philosophy. Structured learning supports the cultivation of essential behaviours for learners to operate at their best rehearsed cognitive potentials in terms of constructive reasoning, visualising problems, and facilitating problem-solving (Tyler et al., 2019). Strong cognitive focus ensures improved quality of sketching information as learners practise and grow in confidence in hand-eye and hand-eye-brain coordination. The cognitive flexing of the brain and psychomotor skills necessary to perform sketching operations is needed in manual sketching, producing an all-rounded individual who can quickly sketch and think creatively in the design process.

The University of Botswana approach adopts an experiential learning disposition to teaching and learning design sketching. The building blocks interwoven in foundation courses articulate the structure of the sketching philosophy in a highly practical way. Defeating inhibition to sketch should be the first step, in the same way the brainstorming approaches work, where there is no right or wrong during the evaluation. The ability of students to visualise should be a confidence builder, which should be carried into the technical design sketching manual and digital platforms. A combination of design fundamentals in terms of the creative process and tools, together with elements and principles of the design, lays a good foundation. Artful thinking should then be a bridge between elements and principles of design and technical design sketching. Once there has been enough practice in the psychomotor skills involved in design sketching, digital tools can be introduced. Digital tools would not be inhibitive to creative thinking at this stage. More activities between manual sketching and digitisation of sketching should be done to expand understanding of how to use both manual and digital sketches (a hybrid approach) (Rapitsenyane et al., 2018). This philosophy has been visualised in Figure 8.12.

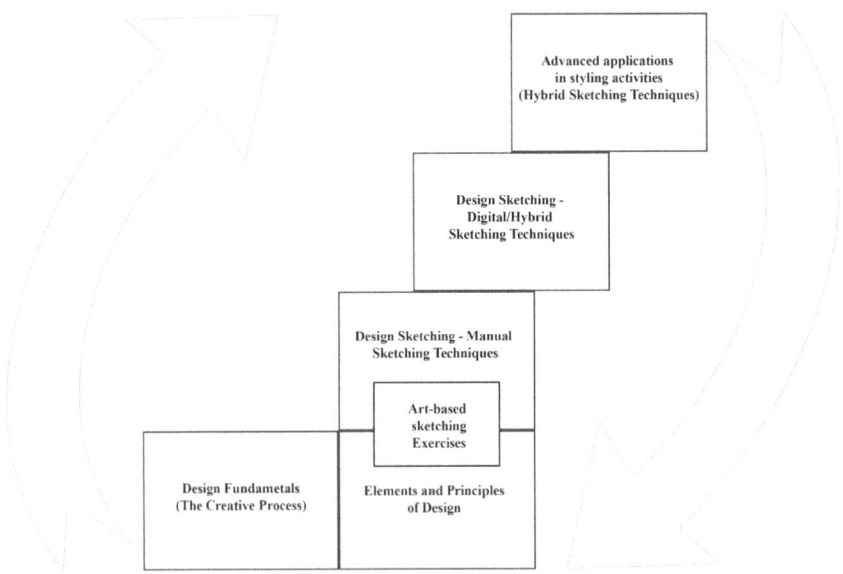

FIGURE 8.12   The University of Botswana design sketching philosophy

Conclusion

There are many approaches to design sketching from professional design practitioners which can be looked at for inspiration. The experience of these professional designers has matured their sketching skills to a threshold of practice such that the skill is now a competence. This competency can be reached when there is an organised way to learn. Exploring ideas needs intuition but communicating them needs the use of the right tools in a proper way. Design educators training students to become influential designers in sketching should systematically impart the right skills. A form requiring a sketching complex forms approach may only be adequately communicated if the student knows or has learned it correctly. Imparting sketching behaviours requires constant practice and should be deliberately built into the learning experience.

The Eurocentric approach to design sketching is documented in many design sketching books. How design educators introduce sketching to students should consider reducing sketching inhibition and making sketching feasible for all students to benefit their design careers. Design sketching remains a critical design skill which supports many design activities in the creative process and therefore requires proper instruction. The chapter's design sketching philosophy has been a well-articulated structured approach.

The following recommendations are made for further exploration of teaching and learning design sketching:

- Design schools should not only look for inspiration in their context for design ideas for solutions but also for pedagogical approaches. Using the context and objects in the context helps students relate learning with the environment and objects in the environment they are familiar with.
- From the sample students' work and classroom activities, the sketching media should be maintained as non-erasable ink. This media type helps students aim for accuracy with every stroke of line they make on paper with no room for error and the flexibility to use an eraser.
- Assessment for learning is vital in design sketching for students to get more involved and practise the skill using feedback from their previous attempts.
- Artful thinking should be explored in design schools to support introduction to design sketching and creative thinking in the design process.
- This design sketching philosophy should be tested in other Afrikan universities offering design programmes for further improvement.

Reference List

Barahal, S. L. (2008). Thinking about thinking. *Phi Delta Kappan*, 90(4), pp. 298–302.

Bilda, Z. A. F. E. R., & Gero, J. S. (2005). Does sketching off-load visuo-spatial working memory. *Studying Designers*, 5(2005), pp. 145–160.

Brun, J., Le Masson, P., & Weil, B. (2016). Designing with sketches: The generative effects of knowledge preordering. *Design Science Journal*, 2. http://doi.org/10.1017/dsj.2016.13.

Chmela-Jones, K. A. (2017). Flourishing in graphic design education: Incorporating Ubuntu as a curricular strategy. *The Design Journal*, 20(sup1), pp. S1048–S1057. http://doi.org/10.1080/14606925.2017.1353048.

Coley, F., Houseman, O., & Roy, R. (2007). An introduction to capturing and understanding the cognitive behaviour of design engineers. *Journal of Engineering Design*, 18(4), pp. 311–325.

Cross, N. (2000). *Strategies for Product Design*, 3rd ed. Milton Keynes: The Open University.

Cross, N. (2006). *Designerly Ways of Knowing*. London: Springer.

Détienne, F., Boujut, J. F., & Hohmann, B. (2004, May). Characterization of collaborative design and interaction management activities in a distant engineering design situation. *6th International Conference on the Design of Cooperative Systems-COOP'04*, pp. 83–98, IOS Press, Amsterdam.

Dulić, O., & Krklješ, M. (2018). A note on the role of drawings in architectural design and education. *Zbornik radova Građevinskog fakulteta*, 34. http://doi.org/10.14415/konferencijaGFS2018.059.

Elsen, C., Häggman, A., Honda, T., & Yang, M. C. (2012, August). Representation in early stage design: An analysis of the influence of sketching and prototyping in design projects. *International Design Engineering Technical Conferences and Computers and Information in Engineering Conference*, Vol. 45066, pp. 737–747, American Society of Mechanical Engineers, Chicago.

Hu, W. L., Booth, J., & Reid, T. (2015, August). Reducing sketch inhibition during concept generation: Psychophysiological evidence of the effect of interventions. *International Design Engineering Technical Conferences and Computers and Information in Engineering Conference*, Vol. 57175, p. V007T06A010, American Society of Mechanical Engineers, Chicago.

Hua, M., Han, J., Wang, P., & Huang, S. (2018). Towards a comprehensive understanding of design sketch – A literature review of design sketch taxonomy and considerations for future research, *Idealogy*, 3(3), pp. 261–277.

Lawson, B. (2002). CAD and creativity: Does the computer really help? *Leonardo*, 35(3), pp. 327–331.

Leclercq, P. (2004). Invisible sketch interface in architectural engineering. *Graphics Recognition. Recent Advances and Perspectives: 5th International Workshop, GREC 2003*, Barcelona, Spain, July 30–31, 2003, Revised Selected Papers 5, pp. 353–363. Berlin, Heidelberg: Springer.

Magoma, C. M. (2016). The Shift and Emphasis towards Curriculum Integration: Meaning and rationale. *African Educational Research Journal*, 4(2), pp. 25–30.

Mangaroo-Pillay, M., & Coetzee, R. (2021). A systematic literature review (SLR) comparing Japanese Lean philosophy and the South African Ubuntu philosophy. *International Journal of Lean Six Sigma*, 13(1), pp. 118–135. https://doi.org/10.1108/IJLSS-11-2019-0118.

Marx, J. (2000). A proposal for alternative methods for teaching digital design. *Automation in Construction*, 9(1), pp. 19–35.

Musta'amal, A. H., Norman, E. W. L., & Hodgson, T. (2009). Observing creative behaviours. In E. Norman & D. Spendlove (Eds.), *The Design and Technology Association International Research Conference 2009*, pp. 60–67. Wellesbourne: The Design and Technology Association.

Oxman, R. (2008). Digital architecture as a challenge for design pedagogy: Theory, knowledge, models and medium. *Design Studies*, 29(2), pp. 99–120.

Plüg, S., & Collins, A. (2022). The work of "authenticity" in the age of mechanical reproduction: Constructions of authenticity in South African artisanal brands. *Journal of Consumer Culture*, 22(2), pp. 417–436.

Ranscombe, C., & Bissett-Johnson, K. (2017). Digital sketch modelling: Integrating digital sketching as a transition between sketching and CAD in industrial design education. *Design and Technology Education*, 22(1), n1.

Rapitsenyane, Y., Moalosi, R., & Mosepedi, T. (2022). A hybrid design sketching approach that can drive critical thinking in design and technology. In *Debates in Design and Technology Education*, pp. 221–237. Abingdon: Routledge.

Rapitsenyane, Y., Moalosi, R., Mosepedi, T., & Sealetsa, O. J. (2019). Critical thinking driven by design sketching in the future. Manual, digital or a hybrid approach? *Proceedings of the 10th International Conference of Science, Maths & Technology Education (SMTE) 2019*, 6–9 November 2019, Réduit, Mauritius.

Rapitsenyane, Y., Ruele, V., Ollyn, M., & Mosepedi, T. (2018). Cognitive impacts of programme revision in teaching design sketching as a modelling tool for industrial design courses at the University of Botswana. *Proceedings of the SADC Conference on Postgraduate Research for Sustainable Development*, 28–31 October, SADC, Gaborone.

Schwartz, D. L. (1995). The emergence of abstract representations in dyad problem solving. *The Journal of the Learning Sciences*, 4(3), pp. 321–354.

Self, D. (2013). *Audio Power Amplifier Design*. Abingdon: Routledge.

Suwa, M., Gero, J. S., & Purcell, T. A. (1998). The roles of sketches in early conceptual design processes. *Proceedings of the Twentieth Annual Conference of the Cognitive Science Society*, pp. 1043–1048, Routledge, Abingdon.

Thurlow, L., & Ford, P. (2018). An Analysis of sketch inhibition within contemporary design education. *Universal Journal of Design Education*, 6(9), pp. 2036–2046.

Thurlow, L., Ford, P., & Hudson, G. (2019). Skirting the sketch: An analysis of sketch inhibition within contemporary design higher education. *International Journal of Art & Design Education*, 38(2), pp. 478–491.

Tovey, M., Porter, S., & Newman, R. (2003). Sketching, concept development and automotive design. *Design studies*, 24(2), pp. 135–153.

Tytler, R., Prain, V., Aranda, G., Ferguson, J., & Gorur, R. (2019). Drawing to reason and learn in science. *Journal of Research in Science Teaching*, 57. http://doi.org/10.1002/tea.21590.

Umeogu, B. (2012). The place of symbols in African philosophy. *Open Journal of Philosophy*, 3, pp. 113–116. http://doi.org/10.4236/ojpp.2013.31A018.

Visser, W. (2006). *The Cognitive Artifacts of Designing*. Florida: CRC Press.

Yang, M. C. (2009). Observations on concept generation and sketching in engineering design. *Research in Engineering Design*, 20, pp. 1–11.

Ziff, M (2020, August 24). *Introduction to Design Process and Programming; Principles of Design*. School of Art and Design, Ohio University. https://people.ohio.edu/ziff/ART%202650/Some%20Principles%20of%20Design%20pdf.pdf.

PART 4

Prototyping Approaches

9

PHYSICAL MODEL MAKING FOR DESIGNERS

Sekao Junior Motshubi, Polokano Sekonopo, Matlhogonolo Letsatsi, and Walter Chipambwa

Introduction

Afrika has many informal sectors which develop many products, such as household furniture, farm equipment, pottery products, and baskets, to name but a few. The development of such products needs a systematic model-making process that must be followed. However, these informal sector designers hardly model their products. They mostly reproduce what is already on the market. They hardly experiment to develop different designs before they settle for one appealing to customers. Prototyping is the only model-making technique that informal sector designers use. These designers never use the lessons learnt from developing these prototypes. In most cases, whatever they have developed is pushed to the market regardless of its flaws. This chapter aims to close this gap and assist designers and students in the creative sector in developing innovative products and services.

Physical modelling is not unique to product design. However, it cuts across almost all design disciplines. Physical models assist in making concepts understood by stakeholders, and design flaws can be identified early when the design is still fluid. In product design, modelling has been practised for a long time. Scholars sometimes interchangeably use the words model and prototype, which brings in some confusion (Burch, 2018). The two mean different things and are used by product designers to communicate different information during the design process. This chapter clarifies this confusion by discussing the following: different types of models, where they are used, materials they are made of, and their purpose in design.

A physical model is a three-dimensional, tangible illustration of a design, system, or service. A model is used to try out processes, establish preferred

DOI: 10.4324/9781003270249-14

design features, and determine if the product will work as expected. Models are experiments that designers do to check if what they have put on paper can be transformed into a product with any challenges. Designers check whether the intended design will do the work before wasting expensive resources on a completed product that might not work or appeal to customers.

Physical Product Modelling

There is scant literature on the Afrikan perspective on product design modelling. The only literature available is a few conference papers and journal articles, which hardly cover model making. Due to the lack of Afrikan literature on this topic, this chapter will have little reference to Afrikan literature. Afrikan children learn product modelling at a young age. They do model making using readily available materials, such as clay, fabric from old clothes, and discarded wires. They use the materials to model items such as dolls, animal toys, and mobility toys, such as cars, lorries, tractors, etc. What children model is influenced by their immediate environment (Setlhatlhanyo et al., 2017). For example, a young boy who sees cattle or looks after cattle will be tempted to model the same from clay. Children always check for proportion, size, and aesthetics as they model such objects. If the modelled toy looks unproportional, it will be destroyed, and the modeller will start afresh. The toy will be kept when the modeller is happy with the outcome. The same applies to making toys using wires. When these models are completed, children usually compete to select the best toy. Such self-taught skills are lost when children go to school, as they are not exposed to product modelling.

In the design process, model making is one of the vital initial steps that can determine the success of the designed product. According to Isa and Liem (2014), the physical model-making process and prototyping are indispensable in the industrial design profession. Through model making, the designer can explore the form, composition, and functionality of the product they are designing. The process of model making is ever-changing, as technology is also changing. Regardless of the time spent making a model, the process allows the designer to test their conceptual ideas that might be wrong (Choi & Zhang, 2015).

The designer might have a good concept on paper, and they might even have good illustrations of the idea, but when a model is made, they quickly show the unforeseen problems of the design. Model making in product design is essential as it gives a visual representation of the design, the appearance of the product, and how it works. Though technology has brought new ways of model making, Kelly (2001) does not support virtual modelling in product design, hence the use of physical models, as these offer more information about the design and related challenges (Hallgrimsson, 2012). The physical model provides better solutions in product design, as the intricate details are clear and workable. Product designers must have sketching, computer-aided design (CAD), and model-making skills. These skills determine the designer's versatility and the kind of products they will produce. Hallgrimsson (2012) argues that physical model making is the standard process even though the computer has proffered other solutions, such as 3D printing. Making a physical model before one makes the final product stimulates the creativity of the designer (Broek et al., 2000). Making a model will identify and correct some of the design dimensions and issues related to usability during the model-making process.

Designers use physical models to envision information about the model's context. It is prevalent for physical models of large objects to be scaled down. Smaller objects can be scaled up for ease of visualisation. The immediate intent of physical modelling is to test aspects of a product against the user requirements. Comprehensive testing of models ensures that a suitable product or service is developed. Physical modelling enables designers to present tangible models to stakeholders. Engaging stakeholders to interact with physical models of products allows designers to attain invaluable feedback that enables them to improve the design and avoid the poor investment of time and resources.

Importance of Model Making in Product Design

Product designers view modelling as essential in designing and producing valuable products. Designers model at different stages of the design process to establish several design-related features that would be difficult to establish without modelling. These features may include shape and form, appearance, proportion, comfort, ergonomics, and stimulation of creativity. It is essential that designers model ideas during designing to establish if the product is doable before the design goes for production (Hallgrimsson, 2012). Models help designers and stakeholders visualise ideas that make design testing and evaluation easy since the design is no longer a flat 2D drawing but 3D and tangible. If something needs to be changed, these changes can be made much earlier during the design stages to make the final product perform optimally (Hallgrimsson, 2012). By modelling, designers can save time, money, and energy (Warfel, 2009). The critical evaluation of models can help designers to change or stop a design which is not viable. Designers use models to demonstrate to their clients what the final product will look like at this early stage. Clients can have input into what is being designed. Thus, models open a dialogue with stakeholders. This is a central tenet in the philosophy of Ubuntu, which is essential to product design, especially when designing products for the Afrikan market. It is through dialogue that designers receive feedback from stakeholders.

In summary, physical models provide critical information about:

- Aesthetic factors, including shape, form, scale, colour, and texture.
- Ergonomics and fit – a model can illustrate how the design will fit the user's body or hand.
- Interactions between the internal and external configuration, for example, the connection of electronic components to the design's external structure.

Model-Making Techniques

There are so many model-making techniques which use a variety of materials and techniques. The section discusses four types of model-making techniques – sketch, block, working model, and prototype – as illustrated in Figure 9.1.

Sketch Model

Sketch modelling uses quick, low-cost modelling techniques to create simple 3D physical models that allow designers to visualise their concepts better than they could by sketching on paper. A sketch model is used at the earliest stages of the design process to establish size, shape and form, proportion, ergonomics, and stability. These models can be made using inexpensive materials that are easy to manipulate or work with; examples of such materials are corrugated cardboard, paper, polystyrene, low-density styrofoam, clay, and wire. The idea is quickly made from 2D into 3D to visualise the design. This low-fidelity model describes the degree to which the model represents the final product. It does not come closer to the appearance of the final product. This modelling technique is done at the very beginning of the modelling process. For example, corrugated cardboard was used to make the model in Figure 9.2 of a kettle. It is easy to get, manipulate to any form or shape, and make it to the required size. Another significant reason for using cardboard is that many different shapes and forms can be made quickly. If the designer wants to discard whatever was made, the designer can do so without incurring any cost since cardboard is freely available from retail shops.

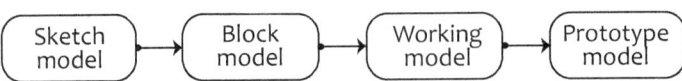

FIGURE 9.1 Model-making techniques

FIGURE 9.2   A kettle sketch model

(Photograph by the authors)

Block Model

Block modelling assists the designer in deciding on the shape, dimensions, and surface details by making 3D models. It can be helpful when specifying the ergonomic factors of the product. Making a series of block models in different shapes and sizes gives designers a feel for the product and its ultimate appearance. A block model is simply a block of what has been designed, showing some critical parts that will make a model look real but not working. With this model, designers want to test ergonomics, appearance, and arrangement of parts, which are done much earlier during the design process. Materials that can be used for this model are the clay, foam, wood, and other materials that are cheap and can be manipulated easily.

This type of model can range from low to middle fidelity, depending on what the designer wants to find out. For example, if the designers want to establish appearance/aesthetic appeal, the block model can be finished to look like the final product. This modelling technique is typically done after the sketch model.

This modelling technique shows clients what the final product will look like and is precisely geared at the aesthetics of the designed ideas. The car model shown in Figure 9.3 is a block model of a side table designed by an industrial design member of staff. The idea was to try the colour combination used in the block model and take it to clients to get feedback on the product's appearance (aesthetic appeal).

FIGURE 9.3   Side table block model

Source: Photograph by the authors

Working Model

A working model is made very close to the final product. If the final product must have a cavity to accommodate parts, outward moving parts like wheels and switches, etc., this model will have such. The only difference from the final product is that this model will not work. This model type is used when the designer wants to make a very close representation of a final product. Materials, parts, positioning, and size of the final product are to be represented in this type of model. The model is a middle fidelity one since it is very close to the final product. This modelling technique is typically done following the block model.

In a working model, the design shows how the moving parts will interact with each other for physical visualisation. The working model represents all the decisions the designers make regarding usability and aesthetics, as these are key to the target customer or client. Figure 9.4 shows a working model of an automated dog feeder.

FIGURE 9.4 Dog feeder working model

Source: Photograph by the authors

Prototype

There are different levels of prototype fidelity, depending on the application and context. The prototype's fidelity is the extent to which the prototype precisely looks like the final product.

- Low-fidelity model is a conceptual representation analogous to a concept. It communicates essential information about form, shape, function, etc. A paper prototype is an excellent example of a low-fidelity prototype.
- Middle-fidelity prototypes communicate more about an idea or concept. They may communicate some, but only some, of the functions of the design.
- High-fidelity prototypes attempt to represent as close as possible the functionality of the final product. They are typically durable to be tested and used by the user group to gather usability data.

A high-fidelity prototype is the last model that designers make to represent the final product after back and forth between the designers and clients, and all stakeholders. This is the first product that will be tested against all parameters that designers, clients, and all stakeholders set in the user specifications. The model represents the first product that looks, feels, and works like the final product. This model will be taken out to be tested in real life to solve the identified problem. All flaws are noted and addressed during testing before the product can be mass-customised or produced. The actual materials, parts, final finish, etc., are used when making this model. Prototypes help the design team discover flaws related to manufacturing the final product. It also allows the design team to learn from the user through user feedback and user trials with the final prototype. The model is high-fidelity since it appears and functions like the actual product. This modelling technique is typically done following a working model.

It must be emphasised that the different modelling techniques described above have been orderly written, from the sketch model down to the prototype. This is the order that must be followed when modelling ideas. However, since the design is iterative, one can always go back to any modelling stage that needs to be beefed up. Alternatively, the designer can decide on any model depending on the required information.

The prototype should have all the design details and must be an accurate representation of the final product (Figure 9.5). A prototype should be made of the same materials as the final product and resemble the final product. According to Choi and Zhang (2015), a prototype can have visual and functional dimensions. The visual fidelity dimension will determine how the product will look when it is finally made, and the functional fidelity will respond to how the product will work. For these two dimensions, designers can test the model in real life, establish its mistakes, and do another prototype until the designer eliminates most of the problems related to the prototype before mass production. Kelly (2001) opines that prototyping makes one understand tangibility and can be done on a service or product. Thus, industrial designers need to use physical models in their design process as they make genuine life products rather than services. Figure 9.4 shows a machine that is used to efficiently extract honey from the honeycombs.

FIGURE 9.5   Honey extractor prototype

Source: Photograph by the authors

Some of the advantages of prototyping include the following:

- Prototyping is another form of exploring design ideas and stimulating creativity.
- Some ideas can be more easily understood in physical form than in 2D drawings.
- Prototypes facilitate communication among design teams and non-technical stakeholders (clients, users, and community).
- It is easy to test a tangible object because users can manipulate it to their satisfaction.
- Prototypes facilitate user trials and testing, e.g., gathering valuable ergonomic data.

Despite the advantages just mentioned, prototypes have some disadvantages, which include the following:

- The creation and iteration of models can be a time-consuming exercise.
- Prototypes can misrepresent the reality of the design context, e.g., a scale model may miscommunicate ergonomic data. Therefore, designers can certainly make assumptions about how precisely a model represents reality.
- The cost of manufacturing prototypes will add to the overall product manufacturing cost.
- Typically, the material used in making the prototype varies from the material used in the final product. Therefore, precise data about the product's functionality or aesthetic qualities may not be collected.
- In the era of sustainability, materials and processes may create waste that affects the environment and humanity.
- Making prototypes demands a level of skill in material processing and technology used to make the prototype.

Conclusion

Physical models have many essential applications in product design, architecture and engineering, medical research, and the automotive industry. They enable designers to identify design flaws during the early stages of the design process and allow the designer to correct them before it is too late and costly. Model designers can show their clients what the finished product will look like, and any suggested changes from the client can be done at minimum or no cost. Through models, designers can establish the feasibility of their designs and make informed decisions on whether to continue with the project or stop it before investing much time and resources. Therefore, Afrikan designers must embrace this design tool (modelling) and use it to help innovate, visualise, and communicate ideas with stakeholders to get their buy-in at the beginning of the product design process. If prototyping is appropriately executed, it can assist Afrikan innovators and designers in producing glocalised products and services. Products that address the local community's needs and equally appeal to the global market. Designers in the informal sector have rudimentary prototyping skills, and there is a need for professional capacity building. Short courses can be conducted to upgrade designers' skills in the informal and creative sectors. Afrikans have abundant skills; what needs to be improved is retooling and upgrading to a professional level. This will allow Afrikan designers to narrate the story from a professional perspective rather than an uncoordinated, informal sector viewpoint.

Reference List

Broek, J. J., Sleijffers, W., Horváth, I., & Lennings, A. F. (2000). Using physical models in design. *Proceedings of CAID/CD'2000 Conference*, pp. 155–163, International Academic Publishers, Lausanne.

Burch, J. (Updated 11 May 2018). *What is the Difference Between a Prototype & A Model?* https://yourbusiness.azcentral.com/difference-between-protoype-model-28781.html.

Choi, Y. M., & Zhang, L. (2015). Student perspectives on fabrication methods and design outcomes. *Archives of Design Research*, 28(4), pp. 49–61.

Hallgrimsson, B. (2012). *Prototyping and Model Making for Product Design*. London: Laurence King Publishing.

Isa, S. S., & Liem, A. (2014). Classifying physical models and prototypes in the design process: A study on the economical and usability impact of adopting models and prototypes in the design process. *DS 77: Proceedings of the DESIGN 2014 13th International Design Conference*, Design Society. https://www.designsociety.org/publication/35348/classifying+physical+models+and+prototypes+in+the+design+process%3a+a+study+on+the+economical+and+usability+impact+of+adopting+models+and+prototypes+in+the+design+process.

Kelley, T. (2001). Prototyping is the shorthand of innovation. *Design Management Journal Summer*, pp. 35–42.

Setlhatlhanyo, K. N., Motshubi, S., & Dichabeng, P. (2017). Improving hands-on experimentation through model making and rapid prototyping: The case of the University of Botswana's industrial design students. *Global Journal of Engineering Education*, 19(3), pp. 219–224.

Warfel, T. Z. (2009). *Prototyping: A Practitioner's Guide*. New York: Rosenfeld Media.

10

TRADITIONAL MANUFACTURING

*Polokano Sekonopo, Sekao Junior Motshubi,
Matlhogonolo Letsatsi, and Samuel
Oluwafemi Adelabu*

Introduction

This chapter discusses the traditional Afrikan practices of processing materials before the Industrial Revolution and globalisation. It seeks to pick those practices which can be regarded as purely Afrikan and tries to compare them with the introduced way of processing materials to find out how the two can be brought together sustainably. The chapter discusses the following materials and processes: leather, ceramics, wood, and metal processing.

Leather

Leather is used in daily life for different purposes. The most widespread use of leather is in shoemaking and other fashion accessories. It makes outstanding quality products, though it is of late being threatened by synthetic leather and other alternative materials. The price of leather products is higher than those made of alternative materials. Therefore, this creates a low demand for leather products compared to cheaper synthetic materials. The leather industry is driven by the agricultural industry, especially the rearing of animals in Afrika.

According to REPOA (2020), global leather exports doubled from US$50 billion in 2001 to US$112 billion in 2014. This shows a need to focus on the leather sector in Afrikan countries because of the booming beef industry. UNIDO (2007a) observes that Ethiopia, which had early contact with Italians, has a more developed leather and leather products industry than the rest of the countries in the region. The need for tannery facilities in Botswana has resulted in the low processing of hides into leather. Despite having three state-of-the art abattoirs, the Botswana Meat Commission has yet to capitalise on its position of having hides as a by-product.

DOI: 10.4324/9781003270249-15

Domestic and wild animals provide leather. However, wild animals tend to be protected, as some are endangered and rarely killed for the leather. Birds, such as an ostriches; reptiles, such as, crocodiles; and wild animals, such as lions, leopards, and cheetahs, produce rare leather for exclusive products. REPOA (2020) posits that global production of hides and skins is dominated by cattle, which accounted for 69% of the total production in 2018, followed by sheep, goats, and buffalo, which represented 15%, 10%, and 7% respectively.

Uses of Leather

Leather is usually used for making personal goods, such as clothes, shoes, hats, purses, belts, sandals, mats, jackets, etc. Wild animals' leather is rarely found, and some have unique cultural uses. For example, leopard skin drapes royalty during their inauguration. It is believed that being draped with the skin of a fierce animal makes one equally fierce and, therefore, able to preside over a large tribe.

Extraction of Leather

Care of the animal before killing it for its meat and the leather enhances the quality of the leather. An animal maltreated during its lifetime and with diseases left without treatment will result in poor leather quality. Animals which have been injured or have scars produce leather that has blemishes, and therefore, it will not be used in a variety of products. Animal skin should be cared for with ointments to heal any wounds. Regular dipping of animals prevents damage to their skin by parasites. In the case of cows, dipping is performed at intervals over several months to kill parasites. To extract excellent leather, the following processes must be observed. First is killing the animal, then skinning, drying, and processing the leather.

Killing: There are procedures for how the animal should be killed and skinned. The animal is firmly tied up to prevent it from haphazardly moving around when it feels pain from the knife or gunshot. This is to prevent it from falling erratically.

Skinning: Improper skinning can result in a wasted piece of leather. The knife to be used should be sharp to aid easy skinning. Care should be taken to prevent blood flow from contacting the animal's skin. This can result in drying or discolouration of the leather. Lack of resources plays a significant role in negatively affecting the quality of the leather.

Hides and skins, which in the past used to be crudely processed with oil tonnage or tanned in pits using vegetable tannins, are still tanned along major rivers such as the Nile and Niger and are used for making simple sandals and other leather articles.

(UNIDO, 2007b, p. 5)

REPOA (2020) report that inadequate slaughter facilities cause animal skin to be mixed with blood and dung, making skin preservation hard.

Drying: The skin is immediately stretched and held around the corners with wooden wedges/pegs. The process utilises sunlight for drying the leather. Salt granules are sprinkled on top of the wet leather to catalyse and speed the drying process. The leather should be dried immediately after skinning, as it is highly susceptible to houseflies' attacks and rotting when wet. The houseflies will lay their eggs on the leather, and the worms will instantly start feeding on it. The leather takes about three to four days to dry thoroughly.

Tanning: Leather production usually involves three distinct phases: "preparation (in the beam house); tanning (in the tan yard); and finishing, including dyeing and surface treatment" (World Bank Group, 1998, p. 404). The two mainly used methods of leather tanning are chrome and vegetable tanning. After drying up, the leather is trimmed to shape, and the removal of the pieces where the stretching-up pegs were hammered takes place. The leather is then dipped in water to soften it before scraping the flesh (defleshing). The leather is then entirely made soft. From this stage, there are two alternatives for leather used; a) it can be used in this state to make products, or b) it can be further processed to remove the fur.

1 The leather used in this state is commonly used for making traditional mats and other products. The leather's aesthetic comes from the fur's colour and texture.
2 Further processing entails treating the leather with chemicals or traditional tree extracts to make the fur fall off. After the fur has been removed, further treatment with colourants is enacted to colour the leather. The leather can be used to make products. Modern tanning involves chemicals and is not as dirty as the traditional way. However, the downside is that using chemicals tends to affect the environment. As many as 90% of the world's leather is chrome tanned, which strongly impacts water pollution (China et al., 2020; Unango et al., 2019; Bacardit et al., 2014).

A newly developed process dubbed Zeology gives hope, as it is chrome-free, heavy metal–free, aldehyde-free, and easy to implement (Smit & Zoon, 2020). Whilst still in its infancy, it remains to be fully commercialised.

Leather Tools and Materials

Traditionally there has been no automation in working leather. The only automation that can be pointed out is when the leather is softened up to make ropes. During this process, leather is cut into thick strips to support a load of about 30 kg made from a huge stone. Several ropes are bundled together and tied up a tree, and the other end will have the stone affixed to act as a weight. A long log will then be pushed across the leather ropes, and the wooden piece will be used to wind the strips and then released to unwind on the weight of the stone. The process is repeated continuously until the ropes are soft enough for various tasks. The other hide-softening process involves applying animal brain tissue mixed with water onto the animal skin. This makes the skin soften up.

Several handheld tools are available for processing leather. The knife, bradawl, needle, and hot iron are tools for processing leather. A knife is a standard tool for working leather. It is used for cutting, scraping, and making incisions on the leather. The bradawl makes holes that the needle will go through when stitching. The hot iron is used for smoothening the leather ends and stitching medium.

Finishing Leather Products

The traditional way of finishing leather products has always been to leave it with its original colour. In some instances, some tree herbs would be ground, mixed with water and applied to the leather as a dye to change its colour. For decorative purposes, beads and other materials can be affixed to the product to make it more appealing.

Challenges Faced by the Leather Industry

Poor infrastructure, including the lack of good roads, inadequate facilities for dipping and veterinary disease controls, and lack of village-level slaughter slabs and drying sheds with water and hygienic conditions, has a significant effect on the quality of the Afrikan raw hides. Putrefaction is a severe problem. Hides and skins that have started to putrefy are not suitable for making products, thus resulting in poor-quality leather (Food and Agriculture Organisation, 1998). UNIDO (2002) concurs that a wide range of factors throughout the leather supply chain contributes to this low level of competitiveness: poor physical infrastructure, low levels of foreign direct investment, inadequate levels of technological development, low productivity, poor quality, inadequate training, lack of working capital, lack of adequate environmental control mechanisms, and factors more directly related to trade and marketing.

Producing footwear and leather products from imported components is feasible for some countries where components can be imported duty-free. However, this operation faces tough competition from countries in the Far East and elsewhere (Food and Agriculture Organisation, 1998).

Prospects for Leather

Leather products are an integral part of Afrikan life, despite the availability of synthetic leather. Afrikan countries are challenged to develop more mature industries for them to be able to process leather and fully produce leather products. This, in turn, gives leeway to cheap imports. The Tanzanian leather industry is on its knees as it is dominated by SMEs, which can only produce 30 pairs of shoes per day due to a lack of resources (REPOA, 2020). Botswana faces similar challenges as locally produced leather products tend to be sub-standard from imports due to a lack of product innovation prowess and machinery. The government of Botswana has long tried to start a leather park where leather value addition will take place. According to McBain (2021), international suppliers depend on raw leather from Afrika to keep up with their orders. This should motivate Afrika to leverage the availability of hides and add value to the materials. Turner (2003) observes that with the passage of the Afrikan Growth and Opportunity Act (AGOA) in May 2000 came many market access opportunities for leather and finished leather products. Over 126 new Harmonized Tariff Schedule categories have been opened to eligible sub-Saharan Afrikan countries for duty-free export to the United States, which include 11 categories in leather, 26 in leather goods, and 89 in footwear for an average tariff savings of 17.71% (Turner, 2003).

Ceramic Manufacturing in Afrika

The art of fired clay or pottery making remains one of the longest-standing Afrikan craft traditions (Willet, 2003). It represents a tangible asset in the definition, interpretation, and reconstruction of Afrikans' socio-cultural history and the ingenuity of the group of people that made and used them. Archaeological evidence of pottery practices across locations in the continent, such as those found in Mali, Ghana, Nigeria, Niger, Chad, Kenya, and Southern Afrika, substantiate that fired clay was prominently among "one of the oldest media and techniques in the history of Afrikan material culture" (Berns, 1989). For instance, in sub-Saharan Afrika, the Aïr region of Niger is home to pottery dating to 10,000 BCE (Haour, 2003). The Nok culture of northern Nigeria has revealed the earliest sculptural terracotta works dated 500 BCE, with recent studies indicating a much earlier date (Breunig, 2014; Fagg, 1990). These and many more findings across Afrika show that the continent has a rich and diverse heritage in making and producing ceramic products, which was and continues to be significant to many aspects of life. The products are used in homes, spiritual settings, and political contexts. The inventiveness and creativity expressed in the works excavated from the various sites strongly indicate that indigenous pottery is an established craft. Potters had already acquired the requisite knowledge and techniques for pottery production.

Despite its age, ceramic production is still alive and evolving in many parts of the continent. From the traditional to the modern age, pottery has supported many communities' industrial, technological, and economic development and promotion. There is no doubt that over the centuries, the craft has witnessed changes in the form, function, and decoration of the products and the shift in manufacturing techniques, the scale of production, and the social status of the potters (Fatuyi, 2015). The recent decades have been particularly significant due to the proliferation of industrial products, socio-economic instability, Western influence, rural-urban migration, and advancement in digital and manufacturing technologies. As a result, native pottery practices are quickly disappearing in most rural communities due to discontinuity in practice and dwindling demand. At the same time, the roles of indigenous potters are gradually being edged out by modern adaptations in ceramic technology. Figures 10.1 and 10.2 show ceramic pots commonly used for storage, especially for grains and water. The water stored in them used to become colder than if stored in other containers. Not only were they functional, but they also had some aesthetic appeal through their finish and symmetrical shape design, which ensured stability when used for storage.

Afrikan Pottery and Technology

Traditional ceramics is a plastic craft produced by forming a clay body into objects of a required shape and heating them to high temperatures in a kiln, which removes all the water from the clay and induces reactions that lead to permanent changes, including increasing their strength, and handling and setting their shape (Robert, 1973). It represents the first synthetic material discovered by man. The final product is attained by subjecting clay forms to sufficiently high temperatures to alter the raw material into objects of permanence and value. Several researchers and scholars in the arts, ceramics, anthropology, ethnography, and archaeology have documented the identities, styles, characteristics, typologies, historical developments, and evolutionary

FIGURE 10.1 Ceramic pot set

Source: Photograph by the authors

FIGURE 10.2 Close-up view of ceramic pot

Source: Photograph by the authors

isolation, oven, and updraft kilns. The first six structures can be placed under the open firing type, while the last two take the form of a wall-enclosed firing. In open firing, pottery vessels are carefully arranged on a bed of fuel on the ground and covered with another layer of fuel on top of the stack. The simple wall-enclosed types in West Afrika resemble the modern-day kiln structure.

Different kinds of combustible materials, including light and heavy fuels, have been used for potter firing by Afrikan potters, depending on their cost and availability within their localities. Although wood, twigs, grass, and many plant wastes are commonly used, dried dung from farm animals such as cows, horses, donkeys, or camels is also used in some areas where these animals are present. Any of these fuels can be used individually or combined in varying proportions. Other procedures, such as the firing location, schedule, and conditions of firing vessels, differ across regions. They largely depend on the disposition of local potters, who consider key socio-cultural and environmental factors for successful outcomes. Without using any heat-measuring device, local potters, in most cases, can intuitively judge the maturing temperature through physical observation.

Since the demand for forest products has increased, leading to deforestation, the felling of trees to serve as a significant fuel source for firing pottery products is already stiffened for the traditional potters. The continuous use of wood for firing at an unsustainable rate, though still relatively cheap, raises environmental issues for the ecological systems.

Post-firing

Unique treatments are done on pottery vessels at the post-firing stage to enhance their physical appearance, strength, and imperviousness to water shortly after they are withdrawn from the fire or kept aside to cool down. These treatments are often done with coating, painting, smearing, smoking, sprinkling, or soaking using a specially prepared organic solution, water, or resin. Coating with a specially prepared organic solution is the most common method, using ingredients from several species of trees. The ingredients are crushed and soaked in water to extract their pigment in hot or cold conditions. Application is made by sprinkling the solution on hot vessels or plunging the vessels in the concoction. Another treatment involves applying resin extracted from some tree species on the inside of the vessel to seal off their pores for waterproofing. Other methods include smearing organic materials on the vessel's surface, smoking by burying hot vessels in organic matter, and plunging in or sprinkling with hot or cold water. Many organic materials vastly used in the post-firing process, as listed in Gosselain (2000, 2002), are meticulously selected. Interestingly, these materials are intricately applicable to other human or domestic uses, such as treating injuries and illnesses, preparing a diet, and dyeing fabrics or baskets. This suggests that the traditional pottery production processes are interwoven into the fabrics of cultural essence and the pillars of the socio-economic and political survival of the Afrikan people.

Challenges Facing the Traditional Ceramic Industry

Pottery is still relevant in both traditional and modern societies. Even though the production rate is declining, traditional pottery is still an indispensable component of indigenous cultures that continues to be valued for its domestic, social, and religious functions. Some types of local potteries are still being sustained by some socio-cultural factors, which include drinking water from storage pots, ceremonial wine pots used to store palm wine during traditional marriages, earthenware pots used for the preparation of traditional medicine, ritual pots/bowls produced by exceptional potters for the adherent of traditional religions, pots produced as gifts and symbols for traditional rites and ceremonies.

There is little doubt that the traditional pottery industry has suffered a significant setback due to a widespread shift in social values and technology (Akinbogun, 2021; Berzock & Frank, 2007; Fatuyi, 2015). Notably, the field of ceramics has witnessed profound changes, including the development of new materials that assume a wide range of applications in modern-day life. These materials are made possible by improving material processing technology, scientific analysis, and understanding of the materials and their behaviour. Production methods have metamorphosed from hand and manual production to high-tech manufacturing processes tending towards a fully autonomous system, e.g., 3D printing. Despite its resilience, the traditional pottery industry is declining in many Afrikan communities. The influence of foreign cultures introduced through colonialism, modern religion, Western education, and industrialisation poses a potential threat to the survival of traditional pottery (Kayamba & Kwesiga, 2016).

Mass-produced containers made from metal, plastic, and enamel serve the same purpose as pottery in a rapidly growing industrialisation and digital revolution era. They are now preferred due to their attractive qualities in weight, form, strength, and fashionable appearance. The social change resulting from globalisation and industrialisation has relegated the craft to the background, making it lose prominence in society. However, several attempts are being made to transform the traditional industry. The lack of extensive documentation on the local histories of many pottery traditions and weak state policies on conserving cultural heritage has endangered the visibility of ageing iconic potters who may have died without due recognition. The threat to sustained local pottery production also hinges on low patronage among an uninterested younger generation. The challenges for the continuity of traditional pottery may further be compounded, as surviving local potters find it challenging to thrive and need help accessing resources that have otherwise become a feedstock for modern industrial processes and rapid urban development.

Wood/Timber

Wood, as provided by plants, has been an integral part of Afrikans, as it has acted as the economic mainstay of nations from time immemorial. The trees that wood came from provided people with food for survival and materials for manufacturing their implements. Artists also used wood to sculpt some art-works for decorative, self-expression, or spiritual purposes. Timber is sourced from living trees that have matured, and these trees undergo some processes which result in solid timber. Traditionally these trees where timber is processed were never planted.

The type of plant/tree and its shape and other characteristics determined what it could be used for. Figure 10.3 shows different oversized cooking utensils used for cooking in big pots, especially during traditional functions where catering is for the masses.

Cooking utensils were made considering the type of pot they will be used on. The designs show that a product is incomplete if some aesthetics are not considered. A closer look at the body of the utensil shows some decorative patterns strategically imprinted into the body using pyrography, i.e., heating a metal piece on an open flame and then marking the wood whilst the metal is still hot. Figure 10.4 shows the pyrographed back part of a cooking utensil.

Family pot cooking utensils were also ergonomically made for the purpose. Different shaped spoons/cooking ladles (Figure 10.5) were available depending on the type of cooking stage, or process one is at, i.e., for stirring and for dishing. Considerations of how the product would be stored after use were also factored in, as the back end of the spoon had a slot for hanging the spoon whilst not in use.

The manufacturing process of the utensils utilised minimal power consumption. Drilling was made with a hot metal piece from an open fire that was being used for other things. Figure 10.6 shows a close-up view of a cooking stirring utensil. The holes were made using a hot metallic rod, and the wires were joined by rolling them over one another.

The dining set (Figure 10.7) was made from a material that is non-toxic and not water-soluble. Ergonomic considerations also played a significant part in how the product was made.

Some other timber by-products are made during wood processing, which could be converted or used for other utility items. Smaller off-cut branches would be made up into smaller storage devices. Other waste by-products (shavings) were left to dry and then used for enhancing open fire for cooking.

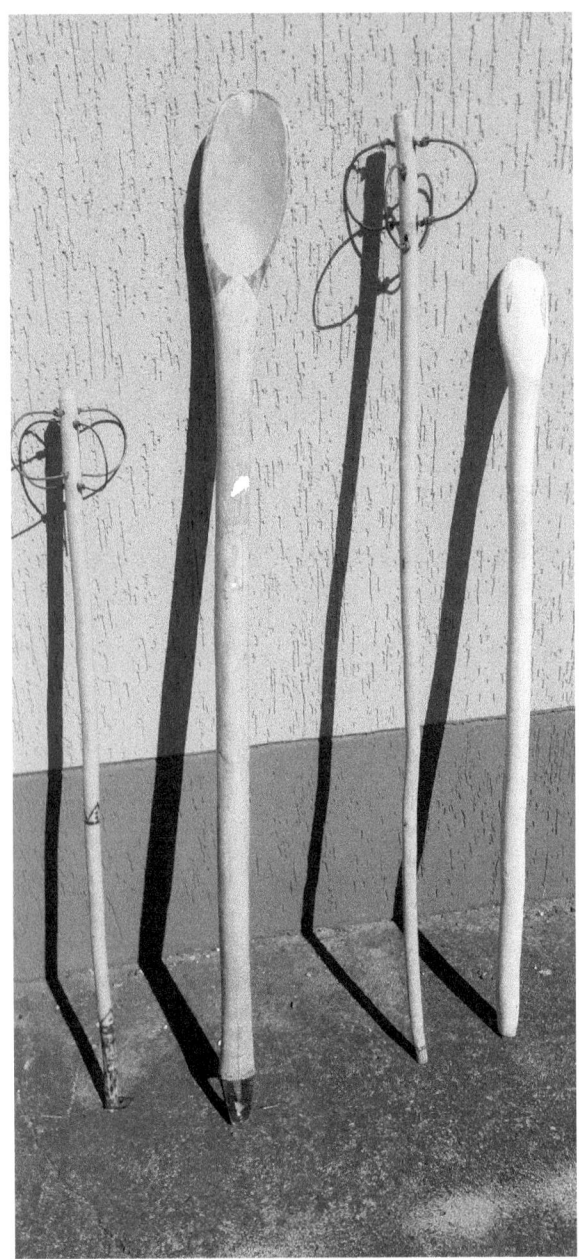

FIGURE 10.3 Traditional cooking utensils

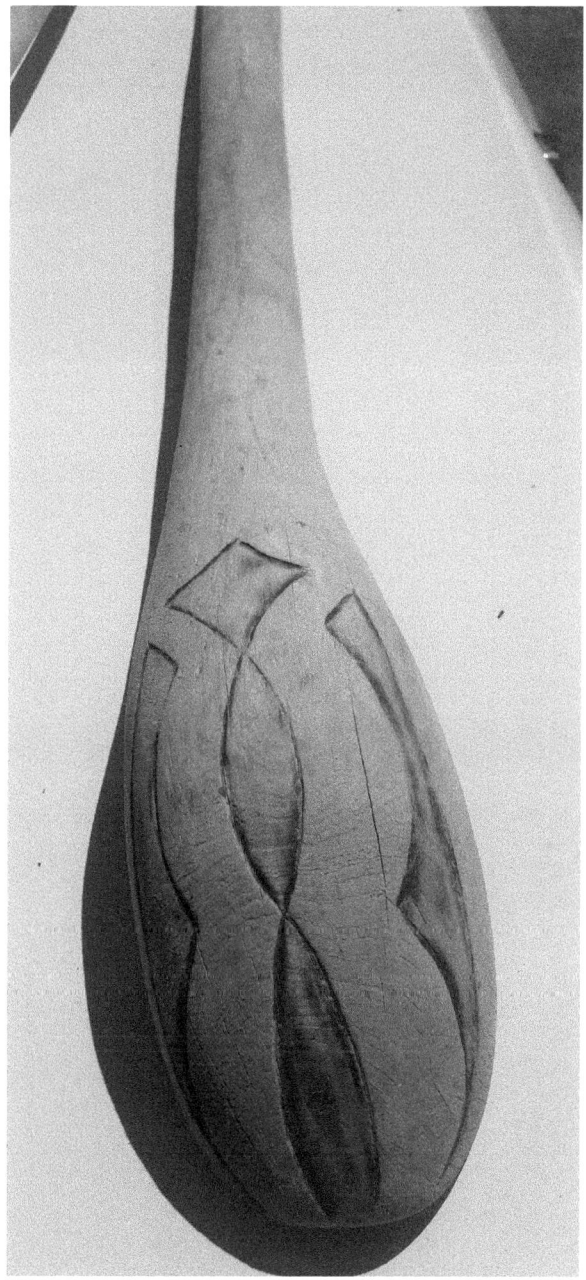

FIGURE 10.4   Back part of a cooking utensil

Source: Photograph by the authors

FIGURE 10.5   Cooking spoon

Source: Photograph by the authors

FIGURE 10.6   Construction of a cooking utensil

Source: Photograph by the authors

FIGURE 10.7 Dining set

Source: Photograph by the authors

Not only were plants providing valuable timber, but they were also providing grass for basketry (see Figures 10.8 and 10.9). Traditional brooms were of different shapes and sizes. The bristle's texture also differs depending on where the product will be used.

The grass and tree bark shells were used for basketry. These were used for various purposes depending on how it was designed and made. Some were used as small handheld carry bags or large storage bags. Figure 10.9 shows flat baskets. These were used for storing small grains and were handy during ploughing as seeds were carried around the field and broadcast from them. Also, during grain processing after harvesting, they will be used for winnowing.

Plant shells were also customised for different uses. Portable water storage was also an integral part of daily living since there was a need to travel from one place to another hunting. Figure 10.10 shows a mobile water storage artefact. It was made from a butternut-like plant which was then left to dry up, after which a hole was made on top and the inside contents were removed. The ropes surrounding the product can also be sprinkled with water to enhance the evaporative cooling of the stored water.

Processing

Plants/wood processing traditionally undergoes two main processes: felling, which is cutting down trees into logs, and seasoning. It also means extracting excess moisture from logs and/or timber before they are used. Softer tree bark was attained by carefully peeling off some trees' outer shells; in others, the rugged bark was first removed, and the inner softer shell peeled into strips.

FIGURE 10.8 Different textures of traditional brooms

Source: Photograph by the authors

FIGURE 10.9 Flat baskets

Source: Photograph by the authors

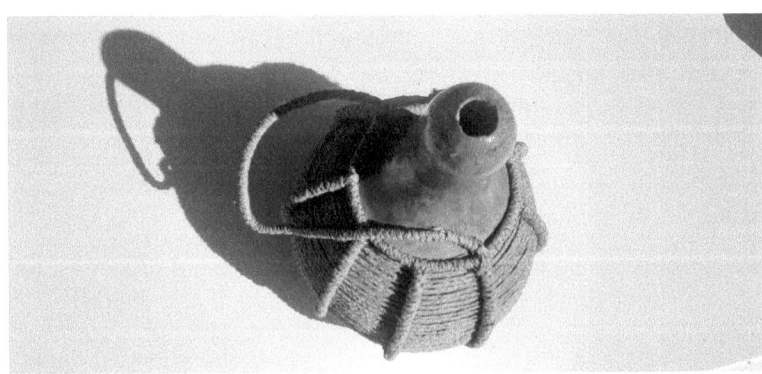

FIGURE 10.10 Mobile water storage

Source: Photograph by the authors

Felling

Axes and small manual saws were traditionally used in this process. The trees to be cut were carefully selected, and the selection was informed by the product for which the tree was to be used. This is still happening among people who use wood to make products. It is important to note that sustainability issues were considered whenever there was a need for a tree to be cut. Some trees were cut during certain times of the year, and even the cutting tools were used selectively. For example, manual saws were not encouraged because it was believed that a tree cut using a saw never grows again. The saw was also used less because the elders realised that a saw cut down trees quickly with less energy, which could have led to cutting down too many trees at once. Using an axe requires a lot of energy. Therefore, its usage will only mean that the tree needed for immediate use will be the only one that will be cut down. To further promote sustainable use of trees/wood, children were only allowed to use these cutting tools (axes and saws) with the elders' permission to avoid indiscriminate cutting down of trees. Elders were responsible for ensuring that no child cut trees indiscriminately.

Seasoning

Before logs could be used, they were allowed to dry up. These logs were not debarked but instead laid down on the ground, which would have been carefully selected for this purpose and given a couple of days to lose some moisture. These logs were turned in every second day to facilitate even drying. For example, if only one log was to be seasoned and used for making an axe handle, the un-debarked log would be kept in a used animal kraal where the log would be covered by dry animal manure. Dry animal waste and urine minimised drying by preventing the development of end split and checks, which might affect the quality of the log.

Traditionally there were few fabrication methods since wood was used to make just a few products. Only a few tools were used, and these tools were informed by what needed to be done. A unique tool will be made for any unique or one-off application.

Traditional woodworking tools were very few. Modern woodworking tools were designed to be used for specific purposes. Wood drilling or making holes in wood was traditionally done when building grass-thatched houses and chairs. Tools used for drilling were a ratchet brace and manual auger, and those who did not have these tools made a hole using a hot wire heated in an open fire. Traditionally, there were few wood joints because only a few wooden products used timber. Instead, products were made from logs that were usually naturally shaped to provide the required shape. The most common joint used was the mortice and tenon joint, reinforced with a dowel and then tied down with a flexible tree bark.

No finish was applied to products made of wood. Pattern-making on some products was the only way products were finished. Currently, most wood carvers still need to apply a finish to their products. They leave them to expose the natural beauty of the wood. For a fine surface finish, products were scraped to take off carving tool marks to achieve a smooth surface. Different types of wood were selected for the specific purpose for which the wooden product would be put. For example, a specific tree with a flexible (elastic or malleable) stock was picked to make a sledge or animal whip. The malleability was to let the stock bend without breaking when the product was in use. Rasps and a very sharp knife or any specially made sharp blade were used on wooden surfaces. From the scientific analysis of the wood from Egyptian tombs, scientists picked the materials and types of joints used (Park, 2016). They further observed that cedar was primarily reserved for the coffins of high-ranking people and noted that different parts of a coffin could use different timbers, depending on their suitability.

Metallurgy in Afrika

The discovery of metal signalled a turning point in people's lives worldwide. The production of sturdy products made of metal was undoubtedly a game changer. As long-lasting and effective products were made, archaeologists provided insights into where and when metal smelting was initially discovered. Different continents at different times showed signs of metalworking. Afrika had some positive ground with smelting.

> The earliest securely dated iron-smelting furnaces in sub-Saharan Africa (ca. 400–200 BCE) were shaft furnaces with multiple bellows and internal diameters between 31–47 inches. Contemporary Iron Age furnaces in Europe (La Tène) were different: the furnaces had a single set of bellows and had internal diameters between 14–26 inches. From this beginning, African metallurgists developed an astonishing range of furnaces, both minor and significant, from tiny slag-pit furnaces in Senegal, 400–600 cal CE, to 21 ft tall natural draft furnaces in 20th century West Africa. Most were permanent, but some used a portable shaft that could be moved, and others used no shaft.
>
> *(Hirst, 2020)*

This confirms that Afrikans had developed better smelting techniques optimised to the conditions under which they will be used.

> Killick suggests that the wide variety of bloomery furnaces in Africa resulted from adaptation to environmental circumstances. Some processes

were built to be fuel-efficient where timber was scarce, and some were built to be labour-efficient where people with time to tend a furnace were scarce. In addition, the metallurgists adjusted their processes according to the quality of the available metal ore.

(Hirst, 2020)

"The earliest records of bloomery-type furnaces in East Afrika were discoveries of smelted iron and carbon in Nubia in ancient Sudan dated at least to the 7th to the 6th century BC" (Metal Casting Institute, 2022, p. xx). They further note, "The site of Gbabiri, in the Central Afrikan Republic, has also yielded evidence of iron metallurgy, from a reduction furnace and blacksmith workshop; with earliest dates of 896–773 BC and 907–796 BC respectively." This confirms that Afrika indeed had some metallurgy in those times. Afrikans used the bloomery process earlier to smelt iron (Hirst, 2020). The bloomery furnace was and is currently being used as a specialised bread oven in a modified form. An extended account of metallurgy in Botswana can be observed from Thebe et al. (2016), alluding that several early metallurgy indicators are in the Tswapong region.

The carbon-dating data from the Tauranga region in West Afrika shows that they have been producing iron since the 4th century (Jemkur, 2004). Whilst sub-Saharan Afrika is thought not to have had ancient iron smelters, Van Noten and Raymaekers (1988) argue that the tall iron bloomers had startled the Cameroonians in the 1950s when digging for sand were produced by immigrants who spoke the languages of the Bantu family and had migrated southwards to Southern Afrika.

In the history of Botswana, metal has yet to be wildly processed from raw to pure metal. Available records show some ancient metallurgy activity in the Tswapong region.

Archaeo-metallurgical research in the Tswapong Hills of southeastern Botswana has yielded evidence for two smelting traditions. Forced draft furnaces were made by Happy Rest and Diamant communities (Kalundu Tradition), while Zhizo communities (Nkope Branch of Urewe) used natural draft furnaces.

(Thebe et al., 2016)

To separate the impurity, copper and nickel are mined and treated through a smelter process. Afterwards, the ingots are sent abroad for final processing into pure copper and nickel metals. A common material, such as steel, is ordered outside the country from the Republic of South Afrika. Of late, iron ore has started to be mined in Botswana and then exported for further processing.

Common Fabrication Metal Processes

In the group of steels, mild steel is the most used. Artisans made different products using hand forging and welding processes. Sheet metalwork is also another standard process used to make products. This process is complemented by soldering, riveting, brazing, and welding. For example, during the manufacture of the cartwheel ring, the metal was heated to the aforementioned recrystallisation temperature (red hot) and then rolled to the desired diameter. Forging was done to join the ends to make a complete ring. The ring ends were heated in charcoal through a continuous air blow using a leveraged hand-operated blower (Mouba). The blower kept the flame constant, hence consistency in temperature. When the overlapped ends were at a red-hot temperature, they were forged by hammering them together. The process was be repeated until the lapped ends formed a homogenous joint. The outer expanded portion was gently hammered to the required width. Sometimes holes through the metals are needed for pinning, decorating, or providing water outlets. Making holes on thick bars also requires excessive heating of metal. While the metal is still red hot, a cold rod of the desired hole will be hammered through the material. This is a repetitive process until the hole is acquired. Finishing the hole to the required size is usually done by hand filing.

Joining together a metal ring with a wooden wheel block was a challenge because the process was done while the ring was still hot. The blocks of wood in a circular form were joined together with strips of wood. To make a solid and firm complete cartwheel (ox-wagon), a ring was to be tight-fitting with a wooden circular shape. Therefore, the metal ring must be heated within the burning charcoal. While the rim/ring was still hot, it was quickly hammered around the rim and immediately cooled with water to avoid burning the formed wooden circular wheel.

Sustainability

The wheel's life span was satisfactory. It was firm enough for a long time in rough road conditions. The cart was used on muddy, stony, and bushed roads but always succeeded. In case of failure by getting loose, some pegging was done and watered to allow expansion for a tight fit. The manufacturing of carts brought sustainable life to manufacturers by getting income from the public. Transport and movement of bulk luggage were made more accessible. The limitations were possible because of the long manufacturing period due to insufficient tooling. The process demanded hard labour. They required modern tools to speed up the manufacturing process. The excellent surface finish was only to a filing process and painting and/or colouring by traditional squeezed gum from trees.

Conclusion

Afrika is blessed with an abundance of human and untapped material resources. Prior to Western civilisation, Afrika had similar or related processes. After colonisation, specific processes were lost due to the adoption of refined processes by the colonisers who had undergone the industrial revolution. Afrikan stolen art pieces by the Western invaders bears testimony that their artefacts were priceless and of good quality. The need for codified ways of sharing knowledge from expert artisans to artisans led to the loss of skills across generations. The ingenuity of Afrikans has been shown by the type of furnaces they developed. It can be concluded that the demise of metallurgy in Afrika has been caused by the slow pace of development or the failure to undergo the Industrial Revolution, which led to other nations perfecting their crafts, leading to mass production. This, in turn, led to Afrikans buying processed materials and finished goods from other nations and thus losing their craft.

Limited development of traditional ways of working materials favouring the colonisers resulted in the demise of the traditional Afrikan way. For example, several pottery traditions face numerous fundamental challenges. To optimise the utilisation of the existing pottery styles and evolve a new style to sustain it and meet modern challenges, several steps can be taken to turn traditional and contemporary pottery around. The modern advances in science and technology should be harnessed to enhance traditional technologies of the people and, by so doing, make them relevant to the social reality and needs of the populace and improve the quality of life, e.g., the design and development of ceramic water filters using traditional materials. More significant improvements in the conceptualisation and visual appeal of traditional pottery products can be further developed for a modern lifestyle.

Traditional products can also be enhanced through thorough quality control and advanced research to redefine their desirable characteristics. Advances in the local technology used by the local potters will aid more productive efforts, reduce stress, and promote better economic activities. Indeed, a cultural-sensitive and artisan-centred design and development approach will prove more effective in adapting modern technologies to traditional pottery practices. Such should include integrating the Fourth Industrial Revolution technologies into the traditional manufacturing processes and enhancing firing and post-firing applications using energy-efficient and less-polluting methods. Capturing the artistry and skills of ageing master potters using digital and media technologies can also help sustain indigenous crafts and inspire a new generation of artisans (Kashim et al., 2013). An early attempt has been made to groom traditional potters to develop the craft with a touch of modernity. A good example was the re-training of traditional potters at Abuja Pottery Training Centre by renowned British potter Michael Cardew (Cardew, 1969).

Through his effort, renowned potter Ladi Kwali was discovered, whose classic works remain treasured today.

The traditional way of leather processing is limited in terms of leather quality. On the other hand, modern production methods give excellent quality leather of varying characteristics. The leather quality is controlled and repeatable, and the production methods also can be made to reduce emissions. There are also deliberate efforts to constantly develop the modern way, as seen with new technologies. The newly developed zoology tanning process is a game changer as it contributes to sustainable leather tanning, thus contributing to the United Nations Sustainable Development Goals. Afrika needs to tap into its vast resources of leather in its locality and beneficiate them by charging a premium price for it, which in turn shall contribute to government coffers and help uplift lives in the continent. The traditional Afrikan way should have been developed and continued alongside the modern way. This will have enabled quicker industrialisation of the continent as there will be less need for expensive machinery from the developed world. The sustainability ideals practised long ago, whereby there was controlled extraction and processing of materials, could still be revived to address environmental issues such as climate change.

Additive manufacturing has revolutionised manufacturing, especially 3D printing, in metal and clay. It will be worth exploring how this technology can be integrated with indigenous knowledge systems and technologies to produce authentic, innovative products which portray Afrikan culture. This is a new research area which needs further exploring.

References List

Akinbogun, T. L. (2009). Anglo-nigeria studio pottery culture: A differential factor in studio pottery practice between Northern and Southern Nigeria. *The International Journal of the Arts in Society*, 3(5), pp. 87–96.

Akinbogun, T. L. (2021). Abundant resources, waning growth; The paradox of ceramic industry in Nigeria. *An Inaugural Lecture Series*, 122(122).

Bacardit, A., Burgh, S., Armengol, J., & Ollé, L. (2014). Evaluation of a new environment-friendly tanning process, *Journal of Cleaner Production*, 65(2014), pp. 568–573. https://doi.org/10.1016/j.jclepro.2013.09.052.

Barley, N. (1994). *Smashing Pots. Feats of Clay from Africa*. London: The British Museum Press.

Berns, M. C. (1989). Ceramic arts in Africa. *African Arts*, 22(2), pp. 32–102. https://doi.org/10.2307/3336716.

Berns, M. C. (1993). Art, history, and gender: Women and clay in West Africa. *The African Archaeological Review*, 11, pp. 129–148.

Berzock, K. B., & Frank, B. E. (2007). Ceramic arts in Africa. *African Arts*, 40(1), pp. 10–17. https://doi.org/10.1162/afar.2007.40.1.10.

Breunig, P. (2014). *Nok: African Sculpture in Archaeological Context* (Peter Breunig, ed.). Africa Magna Verlag. https://books.google.com/books?id=BBn1BQAAQBAJ.

Cardew, M. (1969). *Pioneer Pottery*. London: Longman Group Ltd.

China, C., Maguta, M. M., Nyandoro, S. S., Hilonga, A., Kanth, S. V., & Njau, K. N. (2020). Alternative tanning technologies and their suitability in curbing environmental pollution from the leather industry. *A Comprehensive Review*, April 2020. http://doi.org/10.1016/j.chemosphere.2020.126804.

Daszkiewicz, M., & Maritan, L. (2016). Experimental firing and re-firing. In *The Oxford Handbook of Archaeological Ceramic Analysis*. Oxford: Oxford University Press. https://doi.org/10.1093/oxfordhb/9780199681532.013.27.

Fagg, B. (1990). *Nok Terracottas*. Lagos: National Commission for Museums and Monuments.

Fatuyi, O. A. (2015). *Changes in Technology and Styles of Pottery in South-Western Nigeria*. Ogbomoso: Ladoke Akintola University of Technology.

Food and Agriculture Organisation. (1998). *Development of the Hides, Skins and Leather Sector in Africa*, Report, 9–11 November, Cape Town South Africa. www.fao.org/unfao/Bodies/CCP/hs/98/w9700e.h.tm.

Gosselain, O. P. (1999). In pots we trust. The processing of clay and symbols in sub-Saharan Africa. *Journal of Material Culture*, 4(2), pp. 205–230.

Gosselain, O. P. (2000). Materialising identities: An African perspective. *Journal of Archaeological Method and Theory*, 7(3), pp. 187–217.

Gosselain, O. P. (2002). *Poteries du Cameroun méridional*. Styles techniques et rapports à l'identité. Monographies Du CRA, 26.

Gosselain, O. P. (2008). Ceramics in Africa. In H. Selin (Ed.), *Encyclopaedia of the History of Science, Technology, and Medicine in Non-Western Cultures*, pp. 464–476. Heidelberg, Netherlands: Springer. https://doi.org/10.1007/978-1-4020-4425-0_8911.

Haour, A. C. (2003). One hundred years of Archaeology in Niger. *Journal of World Prehistory*, 17(2), pp. 181–234.

Hirst, K. K. (2020). *African Iron Age – 1,000 Years of African Kingdoms*. www.thoughtco.com/african-iron-age-169432.

Jemkur, J. F. (2004). The beginnings of iron metallurgy in West Africa. In H. Bocoum (Ed.), *The Origins of Metallurgy in Africa*, pp. 33–42. Paris: UNESCO Publishing.

Kashim, I. B., Adelabu, O. S., Fatuyi, O. A., & Fadairo, O. O. (2013). Bridging the gap: Artistry of Felicia Adepelu, Poter of Igbara-Odo, Ekiti State, Nigeria. Critical interventions. *Journal of African Art History and Visual Culture*, 11, pp. 50–64.

Kayamba, W. K., & Kwesiga, P. (2016). The role of pottery production in development: A case study of the Ankole region in Western Uganda. *Net Journal of Social Sciences*, 4(4), pp. 81–90.

McBain, W. (2021). Africa's tanneries go hell for leather to supply new markets. *Magazine, African Business*. https://african.business/2021/07/agribusiness-manufacturing/africas-tanneries-go-hell-for-leather-to-supply-new-markets/.

Metal Casting Institute. (2022). *Iron History*. https://metalcastinginstitute.com/.

Park, A. (2016). *Carving with Stories*. www.carvings-with-stories.co.uk/2016/05/20/ancient-egyptian-woodcarving/.

REPOA. (2020). *Tanzanian Leather Chain*. https://media.africaportal.org/documents/Tanzanias-leather-value-chain-review-final.pdf.

Robert, F. (1973). *Practical Dictionary for Pottery*. Lanham: Pitman Publishing.

Smith, A. L. (2001). Bonfire II: The return of pottery firing temperatures. *Journal of Archaeological Science*, 28, pp. 991–1003. http://doi.org/10.1006/jasc.2001.0713.

Smit & Zoon (2020). *Zoology, the Sustainable Tanning Concept*. www.neratanning.com/zeology/?gclid=Cj0KCQiAr5iQBhCsARIsAPcwROMz7u5WIYW9TKqbAf-fnBclwI3wBudod2afgKnkZjruY-6aMRwOY4kaAmuoEALw_wcB.

Thebe, P. C., Huffman, T. N., Watkeys, M. K., & Tarduno, J. A. (2016). Ancient metallurgy in the Tswapong Hills, Botswana: A preliminary report on archaeological context. *Southern African Humanities*, 28(1), pp. 119–152.

Turner, S. (2003). *Regional Leather, Leather Goods and Footwear Supply Chain Study for Southern Africa*, Technical Report, Gaborone Botswana. https://pdf.usaid.gov/pdf_docs/Pnacw210.pdf.

Unango, F. J., Duraisamy, R., & Ramasamy, K. M. (2019). A review of eco-friendly preservative and bio-tannin materials using powdered barks of local plants for the processing of goatskin. *International Research Journal of Science and Technology*, 1(1), pp. 13–20.

UNIDO. (2002). *A Blueprint for the African Leather Industry*. https://leatherpanel.org/sites/default/files/publications-attachments/a_blueprint_for_the_african_leather_industry.pdf.

UNIDO. (2007a). *Present and Future Role of Africa in the World Leather and Derived Products Industry and Trade, Sixteenth Session of the Leather and Leather Products Industry Panel*. https://leatherpanel.org/content/present-and-future-role-africa-world-leather-and-derived-products-industry-and-trade.

UNIDO. (2007b). *Present and Future Role of Africa in World Leather Trade, 16th UNIDO Leather Panel Meeting*, May 2007, Brazil. https://leatherpanel.org/sites/default/files/publications-attachments/present_and_future_role_of_africa_in_the_world_leather_and_derived_products_industry.pdf.

Van Noten, F., & Raymaekers, J. (1988). Early iron smelting in central Africa. *Scientific American*, 258(6), pp. 104–111.

Willet, F. (2003). *African Art*, 3rd ed. London: Thames & Hudson.

World Bank Group. (1998). *Pollution Prevention and Abatement Handbook*, Report, July 1998. www.ifc.org/wps/wcm/connect/18a02881-a68d-4d65-abd0-8ec234c7ea9b/tanning_PPAH.pdf?MOD=AJPERES&CVID=jqeDjDv.

11

ADDITIVE MANUFACTURING IN AFRIKA

Victor Ruele

Introduction

Additive manufacturing, popularly known as 3D printing, uses digital technology to manufacture products that exhibit high accuracy and complexity, an improved lead time, ease of manipulation for resizing, and the elimination of the need for tooling. It also allows for the customisation of products. All these attributes make AM a niche opportunity to effect industrialisation in Afrika. Wu et al. (2017, p. 2) describe AM as "an umbrella term for techniques in which three-dimensional objects are built from sequential layers of material." Similarly, Deckers et al. (2014, p. 24) define AM as "the process of joining materials to make objects from three-dimensional (3D) model data, usually layer by layer, as opposed to subtractive manufacturing and formative manufacturing methodologies." According to Baumers and Holweg (2019), the primary method of production for the AM process is the layer-by-layer deposition of material in a geometrically defined, three-dimensional space. This deviates from subtractive manufacturing, where components are prefabricated and machined to the desired and designed geometry (Klenam et al., 2022). It is a method that performs rapid prototyping using computer-aided design (CAD). This technology has been used mainly for prototyping during the last three decades (Vayre et al., 2012). However, this technology has evolved in recent years from primarily a prototyping tool to a useful end-product fabrication method in some high-value manufacturing applications (Wu et al., 2017). Wang et al. (2019, p. 24) add that the AM process allows for making components in various materials without the requirement for tooling, assembly lines, or supply chains. AM allows for the design of variegated geometries and shapes that would be difficult or impossible to attain using

DOI: 10.4324/9781003270249-16

traditional manufacturing methods. It also allows the creation of customised and one-off products on demand, reducing the need for extensive inventories. Klenam et al. (2022) and Blakey-Milner (2021) argue that AM merits include easy manufacturing, reduced labour, mass customisation, on-demand manufacturing, industrial efficiency, decentralised manufacturing, quality improvement, and component manufacturing.

There are different types of AM technologies, each with its strengths and weaknesses. Some of the most common types include:

1 Fused deposition modelling (FDM): This is the most common type of 3D printing technology used to create objects from thermoplastic materials. The material is heated and extruded through a nozzle and deposited layer by layer to form the object. Figures 11.1–11.3 show some examples of students' work at the University of Botswana, where this technique has been used.
2 Stereolithography (SLA): This technology uses a laser to cure a liquid resin, forming solid layers that build up to create the object.
3 Selective laser sintering (SLS): This technology uses a laser to sinter powdered materials, such as metals and plastics, to form solid layers that build up to create the object.
4 Carbon-fibre-reinforced polymers (CFRP) and glass-fibre-reinforced polymers (GFRP): This technology uses a similar process to the FDM but is reinforced with carbon or glass fibres that make the part stronger.

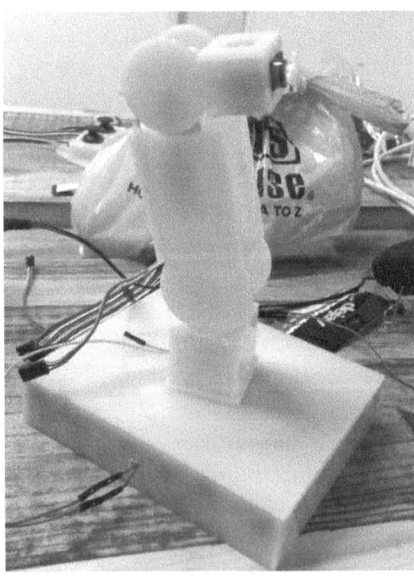

FIGURE 11.1 A robotic arm produced using the FDM method

Source: Image by Lusentfo Nkambule and Yussuf Paul Guesela

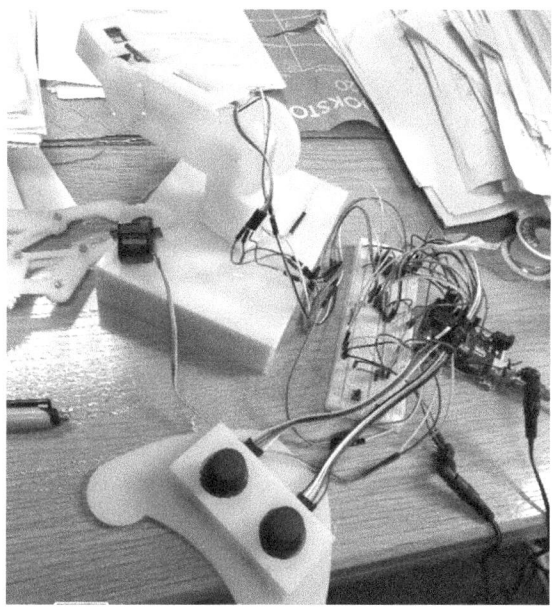

FIGURE 11.2 Testing and assembly of the robotic arm model

Source: Image by Lusentfo Nkambule and Yussuf Paul Guesela

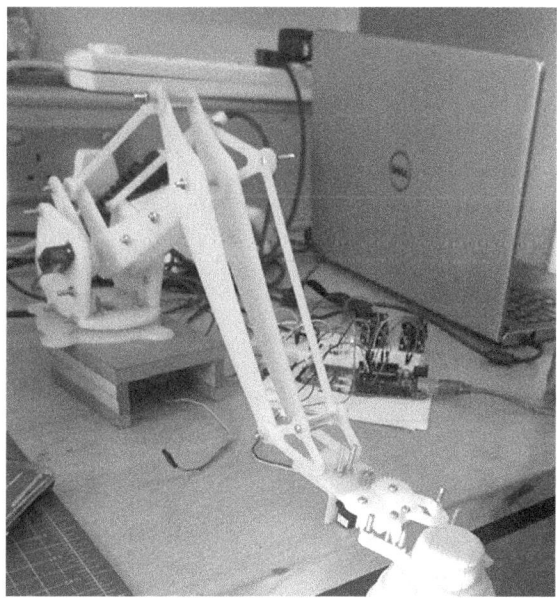

FIGURE 11.3 Assembled robotic arm

Source: Image by Lusentfo Nkambule and Yussuf Paul Guesela

While AM has many advantages, it also has some limitations. The technology is not yet capable of producing large objects or objects with high strength-to-weight ratios, and the materials used are currently limited. Additionally, the cost of AM systems can be prohibitively high for some businesses. Despite these limitations, AM is expected to grow in popularity and importance. As technology advances and new materials become available, AM will likely play an increasingly important role in manufacturing and product development. The following section will review some policies aligned with digital industrialisation in Afrika.

Afrika's Transformation Agenda

In May 2013, Afrikan Heads of State declared the continent's re-dedication towards attaining the Pan-Afrikan Vision of "an integrated, prosperous and peaceful Afrika, driven by its citizens, representing a dynamic force in the international arena" (African Union, 2013). This gave birth to Agenda 2063 on how Afrika intends to achieve this vision within 50 years. Agenda 2063 is Afrika's blueprint and master plan for transforming Afrika into the global powerhouse of the future. Afrika's strategic framework aims to achieve its inclusive and sustainable development goal.

Agenda 2063 is linked to the United Nations Sustainable Development Goals (SDG). For example, Agenda 2063, Goal 2 of a well-educated citizens and skills revolution underpinned by science, technology, and innovation equates to SDG 4 of ensuring inclusive and equitable quality education and promoting lifelong learning opportunities for all. Agenda 2063, Goal 4 of transforming economies equates to SDG 8 to promote sustained, inclusive, and sustainable economic growth and full and productive employment and decent work for all, and SGD 9 to build resilient infrastructures, promote inclusive and sustainable industrialisation, and foster innovation. Agenda 2063, Goal 10 includes building of world-class infrastructures across Afrika.

Amongst some of the goals of Agenda 2063 is the adaption of a development agenda that is focused on renewed economic growth, social progress, the need for people-centred development, information communication technology and infrastructure connectivity, youth empowerment, and emerging development and investment opportunities in areas such as infrastructure development, agri-business, health, and education and the value addition in Afrikan commodities. These are areas in which AM can contribute to being successfully achieved. The Science Technology Innovation Strategy for Afrika (STISA) is one continental framework that addresses science, innovation, and technology issues. The strategy places science, technology, and innovation at the core of Afrika's socio-economic development. It is envisioned that STISA will impact innovation in critical sectors such as the environment, agriculture, energy, health, infrastructure development, security, mining, water, etc.

STISA is based on four pillars: promoting entrepreneurship and innovation, building and/or upgrading research infrastructures, enhancing professional and technical competencies, and providing an enabling environment for science, technology, and innovation development in Afrika. The national development plans of Afrikan countries and the strategic plans of the regional economic communities should focus on science, technology, innovation, and manufacturing-based industrialisation. AM plays a crucial role in innovation and manufacturing-based industrialisation.

Afrika must address the following factors to transform into a digital manufacturing base.

1 Investment in technology: Afrikan governments can prioritise investment in the technology sector and provide funding for purchasing and maintaining AM equipment. This will increase the availability of the technology and make it more accessible to local businesses.
2 Promotion of entrepreneurship: Governments can support entrepreneurs interested in developing AM-based businesses. This can be done through tax incentives, subsidies, and mentorship programmes.
3 Developing human capital: Governments must invest in developing human capital by encouraging training programmes and educational initiatives focusing on AM. This will increase the number of skilled professionals in the field and help businesses adopt the technology.
4 Promoting research and development: Governments can encourage research and development in AM by funding research initiatives and establishing partnerships between universities and businesses.
5 Encouraging international collaborations: Governments can encourage collaborations with countries with advanced technology and AM. This will increase access to the latest technology and knowledge and help local businesses adopt the technology.

These policy directions help address the challenges faced by Afrika in the adoption of AM and other emerging technologies.

The State of Additive Manufacturing in Afrika

Klenam et al. (2022) report that, in Afrika, out of 500 the word is repeated articles published between 2005 and 2021, South Afrika has the highest research throughput, whereas about two-thirds of the continent needs to participate in this burgeoning field actively. The most widely used AM techniques are selective laser melting, fused deposition modelling, and direct energy deposition. The studies reviewed earlier indicate that Afrika lags behind the world in AM and other emerging technologies. There is a need to conduct research and development in AM to raise interest in Afrika. However, most universities and research centres across Africa have purchased 3D printing facilities, and research in the area is expected to increase rapidly.

This is a niche area in which Afrika could leapfrog the manufacturing sector. Thus, Afrika must build collaborative partnerships to become an active global player in AM. The healthcare industry is one of the main areas where AM is highly utilised. It is used in manufacturing medical devices and instruments that are patient-specific and need-based (Popov et al., 2022). Recent developments in the past two years due to the outbreak of the COVID-19 pandemic have catalysed the need to revolutionise the manufacturing sector globally. The World Economic Forum (2019) argues that the COVID-19 pandemic accelerated the need for people to be digitally networked. There was a sudden shift towards digital transformation activities and services. The COVID-19 pandemic has shown the shortcomings of traditional manufacturing practices, which rely on high labour intensity. During the COVID-19 pandemic, AM demonstrated a strategic approach to satisfy the surge of emergency medical supplies at the pandemic's peak. The pandemic disrupted and halted continental supply-chain business; AM technologies enabled the national production of parts using AM facilities in most universities and research centres in Afrika. Manufacturers and research universities were proactive in filling the gaps in healthcare equipment for medical professionals and patients worldwide (Klenam et al., 2022; Bolaane et al., 2022). AM assisted in designing and manufacturing personal protective equipment (PPE), diagnostics and testing, development of medical devices, emergency hospitals and dwellings, and visualisation aids (Wang et al., 2021). The role of AM techniques was critical to this endeavour. Intricate shapes of components were designed with minimal wastage of material and deployed much faster (Klenam et al., 2022). The designs were developed at a low cost and in an environmentally friendly manner. Thus, the advantages of AM processes outweigh the disadvantages and are more disruptive to conventional manufacturing methods.

Opportunities for Additive Manufacturing in Afrika

Medical applications of 3D printing can be categorised into anatomical models, tissue and organ fabrication, customised prosthetics, patient-specific orthopaedic implants, and pharmaceutical research (Kudryavtseva et al., 2020). Several Afrikan countries have been producing medical devices from 3D printing. Innovative 3D-printed, robotic prosthetic limbs provide Afrikan amputees with an affordable and high-quality alternative to conventional prostheses. For example, in South Afrika, amputee care is an expensive service that most people cannot afford. Prosthetics cost between US$20 000 and US$100 000. In South Afrika, Ukuhamba Prosthetics, a small business, uses 3D technology to produce low-cost prosthetic limbs from recycled water bottles. This offset the high expenses incurred from conventional prosthetic limbs. Apart from custom prosthetics, AM can make orthotic devices, surgical guides and templates, tissue engineering and regenerative medicine, dental implants and prosthetics, medical models and simulations, and drug delivery systems.

Furthermore, the University of Botswana, in collaboration with the Central University of Technology in South Afrika and the Botswana Institute of Technology Research and Innovation, are working on a project to produce patient-specific (customised) medical implants. The Southern Afrika Innovation Support programme funded the project through the Finnish Ministry of Foreign Affairs. The project aims to roll it to other countries in sub-Saharan Afrika to promote AM innovation.

Globally, about 900 million people have no access to electricity, and about 565 million (72%) live in sub-Saharan Afrika (WWF, 2023). Approximately 80% of the energy consumed for cooking in sub-Saharan Afrika is generated primarily from wood or charcoal. This is increasing deforestation, greenhouse gas emissions, habitat destruction, and the loss of nature. This makes Afrika the world's largest household black carbon emitter, with households accounting for 60–80% of emissions, with implications for climate change and the health of communities (WWF, 2023). It is envisioned that the overall fossil fuel reserves will be depleted within the next 100 years at the current consumption rate. This requires countries to reduce their over-dependence on carbon-based fossil fuels and introduce renewable energy resources (Wang et al., 2022; Kermavnar, 2021; Stavropoulos, 2018). This is anticipated to reduce greenhouse gas emissions enormously. These ingenuities can be achieved by developing low-cost energy storage and conversion devices using emerging technologies such as 3D printing (Wang et al., 2022).

The adoption of AM technology in the construction industry in Afrika still needs to grow. However, there has yet to be some progress in using technology on a small scale. For example, a small-scale pilot approach for wall buildings has been experimented in Kenya, Morocco, Tunisia, South Afrika, and Egypt. This technique is customised to produce low-cost and sustainable housing and can be extended to other emerging economies. The adoption of AM in the construction industry will enable projects to be carried out time-ously and at a low cost (Sakin & Kiroglu, 2017).

Many countries are reducing their carbon footprint by introducing electric vehicles because the transportation industry increases the greenhouse gas footprint. The move from a fossil-energy-dominated transport industry towards an equitable, clean, and sustainable mobility ecosystem, for example, with electric buses, vans, cars, and motorbikes, is gathering pace. This presents an opportunity to reduce the carbon footprint from vehicle emissions. Electric mobility presents an excellent opportunity for decarbonisation. This is an opportunity to leapfrog into a green economy since the penetration of conventional vehicles is still low. AM will play a leading role in manufacturing electric mobility parts and components. Barclays Research (2021, p. 7) states that "AM is primarily used in the automotive sector for prototyping; small, low-volume batch or customised parts; and production of manufacturing tools."

It has been found that AM technology can be used to produce synthetic diamonds (Thomas & Srinivasan, 2018; Kim et al., 2019; Zhang et al., 2020). For instance, Zhang et al. (2020) highlight the benefits of using AM in diamond production, such as improved precision and accuracy, the ability to produce customised shapes and sizes, and reduced waste. According to them, synthetic diamond is chemically and physically identical to natural diamond. 3D printing technology can create diamonds in custom shapes and sizes, which can be useful in producing jewellery and other specialised applications. Furthermore, Kim et al. (2019) studied the traditional methods of synthetic diamond production and the challenges associated with these methods. They argued that traditional diamond production methods could be time-consuming and labour-intensive, but with 3D printing, they found that the process can be significantly faster and more efficient. The authors concluded that 3D printing technology has the potential to revolutionise the production of synthetic diamonds for industrial applications. However, they suggest that further research is needed to optimise the process and improve the quality of the printed diamonds. Overall, AM technology has the potential to revolutionise the production of synthetic diamonds, making them faster, more efficient, and more customisable.

Another potential application of AM in the ceramic industry involves producing functional components, such as ceramic parts, for use in the aerospace, automotive, and biomedical industries. 3D printing creates complex geometries that are impossible with traditional ceramic manufacturing methods and reduces the need for expensive tooling and moulds. This technological capability can produce these elements with high precision and accuracy, which can be particularly useful for restoration or replication projects. Overall, AM has the potential to significantly benefit the ceramic industry by allowing for greater customisation, precision, and efficiency. As technology advances, we will likely see an even more significant future impact of 3D printing on the ceramic industry (Zhang & Chen, 2016).

AM also presents manufacturing opportunities in the jewellery-making industry. For making jewellery, Afrika is endowed with abundant mineral resources, such as gold, platinum, diamonds, copper, etc. The geometrical freedom AM makes it an attractive technology for use in the jewellery industry. In many cases, layer-wise building significantly reduces waste material, part weight, and several parts in new ways that were impossible before (Gordon et al., 2016). Against this background, AM technologies present opportunities within the jewellery manufacturing sector in Afrika. This will enable Afrikan countries that produce minerals to add value to them instead of exporting raw materials.

Barriers to the Adoption of AM Technologies in the Manufacturing Industries

Despite the ongoing growth and benefits of 3D printing, the uptake of AM technologies in many industrial sectors still needs to be improved. Wu et al. (2017) identified several barriers preventing the widespread adoption of AM in the manufacturing industry.

1 One significant barrier is the high cost of equipment and materials. 3D printers and related equipment can be expensive, and raw materials can also be prohibitive for many developing countries. Additionally, many developing countries need more infrastructure and technical expertise to operate and maintain 3D printing equipment.
2 Lack of access to digital design and modelling software. Many developing countries need more resources and expertise to create digital designs and models that can be used in AM. Access to digital design and modelling software can also limit the ability of small and medium-sized enterprises in emerging economies to innovate and compete in the global marketplace.
3 Lack of regulations and standards for AM. In many developing countries, there are no regulations or standards to govern the use of 3D printing technology. This lack of regulation makes it difficult for businesses and individuals to use and maintain 3D printing equipment properly, creating health and safety concerns.
4 Lack of awareness about the potential benefits of AM among policymakers and the public in developing countries is a barrier to its adoption. Without a clear understanding of the benefits of the technology, it is not easy to secure funding and support for its implementation.

Addressing these barriers will be essential for successfully implementing AM in Afrika.

The Infusion of the Spirit of Ubuntu in Product Design

The philosophy of Ubuntu is particularly relevant to AM, especially in humanising technology. It encourages designers to consider the needs and perspectives of all stakeholders, including users, communities, and the environment. The author argues that by incorporating Ubuntu values into AM, designers can create more inclusive, sustainable, and socially responsible products. For example, a furniture company in South Afrika has adopted Ubuntu principles in its design process, which has led to the creation of products that are not only functional but also culturally meaningful and relevant to the community. The company has successfully built a strong relationship with its customers and the community by providing them with products that reflect their cultural identity and heritage (Zungu & Mkhize, 2016).

First, Ubuntu encourages designers to think about the impact of their products on the community and to design for the common good. This means considering not only the product users but also the broader community that may be affected by it. For example, designing a product that is environmentally friendly or promotes social inclusion can positively impact the community (Ndlovu & Mkhize, 2020).

Second, Ubuntu stresses the importance of collaboration and co-creation in the design process. This means involving users, stakeholders, and other experts in the design process to gather diverse perspectives and to ensure that the final product meets the needs of all parties. This can lead to more innovative and effective solutions.

Third, Ubuntu emphasises the importance of empathy and understanding in the design process (Dlamini & Mthembu, 2019). This means taking time to understand users' and other stakeholders' needs, wants, and perspectives. This results in a deeper appreciation of the problem and more human-centred solutions.

Ubuntu is a philosophy that emphasises the interconnectedness and interdependence of all people and the importance of community and mutual support. In product design, the spirit of Ubuntu can be applied in several ways:

1 Community involvement: Designers can involve the local community in the design process by soliciting feedback and ideas from people using the product. This helps to ensure that the product meets the preferences and needs of the community and helps to build a sense of ownership and engagement (Nkosi & Mkhize, 2015).
2 Collaboration: Designers can work closely with other professionals and stakeholders, such as engineers, manufacturers, and suppliers, to create a product that is not only functional but also sustainable and socially responsible (Dlamini & Mthembu, 2019; Mngomezulu, 2018).
3 Inclusivity: Product design can also be informed by understanding the diverse needs and abilities of different users. This may involve designing products with accessibility features or creating products that are more affordable and accessible to a broader range of people.
4 Empathy: Product design can also be guided by a sense of empathy and understanding of the user's experience. This may involve researching and testing products with users to understand their needs and preferences better.

The spirit of Ubuntu is an approach that encourages designers to consider the needs and perspectives of all stakeholders and create products that are inclusive, sustainable, and socially responsible (Dlamini & Mthembu, 2019; Mngomezulu, 2018). It emphasises the interconnectedness and interdependence of all people and encourages treating others with compassion, respect, and empathy. The spirit of Ubuntu in product design emphasises the importance of community, collaboration, inclusivity, and empathy in the design process. By considering these values, designers can create products that are functional, socially responsible, and beneficial for the community. The infusion of the spirit of Ubuntu in product design offers a unique opportunity to create products that serve their intended purpose and embody the values of community and interconnectedness. By embracing this approach, designers can help to create a more sustainable and harmonious world, one that is built on mutual respect and shared responsibility. This approach to design is particularly relevant in today's globalised world, where the boundaries between cultures are becoming increasingly blurred, and the importance of cultural understanding is growing. Through the infusion of the spirit of Ubuntu in product design, designers can promote greater cultural awareness and help to build bridges between people from different backgrounds.

Conclusion

Digital manufacturing, especially AM, offers unique opportunities for Afrika to address some pressing challenges, e.g., mineral beneficiation, youth unemployment, access to renewable energy, green industrialisation, transition to zero net, etc. AM offers prospects for entrepreneurship for the massive youth population in Afrika and exposes them to a digital world where access to information, innovative business ideas, and employment opportunities are within their grasp. This will go a long way towards alleviating the problem of youth employment in Afrika and other challenges. This demands that the education sector, especially universities and research centres, be at the forefront of curriculum development across the continent and support science, technology, engineering, and mathematics education. Such an approach will groom future scientists, technologists, and engineers.

Moreover, research centres should collaborate with local manufacturing industries to quickly transfer knowledge and technology. This will close the gap where Afrika imports more than 60% of spare parts and machine components, which can easily be 3D printed. Apart from this sector, other industries that can benefit from AM in Afrika include footwear, electronics, medical devices, construction, health, and automotive industries.

AM can promote economic development and job creation within local communities in Afrika. Local businesses and individuals can establish their own AM operations by providing the necessary infrastructure and training, thereby creating new job opportunities and contributing to the local economy. The decentralised and collaborative nature of AM can promote the spirit of Ubuntu by encouraging greater community involvement in the production process. For example, local community workshops could be set up to produce products using AM, promoting greater collaboration, and sharing knowledge and skills. The spirit of Ubuntu can be applied to AM technologies by promoting a more localised and community-driven approach to product design and production. By embracing this approach, AM can help promote more significant economic development and job creation, fostering greater community involvement and collaboration.

Reference List

African Union. (2013). *Agenda 2063*. https://au.int/agenda2063.

Barclays Research. (2021). *Additive Manufacturing: Advancing the 4th Industrial Revolution*. www.cib.barclays/our-insights/3-point-perspective/additive-manufacturing-advancing-the-fourth-industrial-revolution.html.

Baumers, M., & Holweg, M. (2019). On the economics of additive manufacturing: Experimental findings. *Journal of Operations Management*, 65(8), pp. 794–809.

Blakey-Milner, B., Gradl, P., Snedden, G., Brooks, M., Pitot, J., & Lopez, E. (2021). Metal additive manufacturing in aerospace: A review. *Material Design*, 209, p. 110008.

Bolaane, B., Moalosi, R., Rapitsenyane, Y., Kgwadi, M., Kommula, V., & Gandure, J. (2022). A response to the COVID-19 pandemic: Experience of the University of Botswana. *COVID*, 2, pp. 1538–1550. https://doi.org/10.3390/covid2110110.

Deckers, J., Vleugels, J., & Kruth, J. P. (2014). Additive manufacturing of ceramics: A review. *Journal of Ceramic Science and Technology*, 5, pp. 245–260.

Dlamini, L., & Mthembu, S. (2019). Designing for Ubuntu: A human-centered approach to product design. *Journal of Design and Technology Education*, 14(1), pp. 23–30.

Gordon, E. R., Shokrani, A., Flynn, J. M., Goguelin, S., Barclay, J., & Dhokia, V. (2016, April). A surface modification decision tree to influence design in additive manufacturing. In *International Conference on Sustainable Design and Manufacturing*, pp. 423–434. Cham: Springer.

Kermavnar, T., Shannon, A., & O'Sullivan, L. W. (2021). The application of additive manufacturing/3D printing in ergonomic aspects of product design: A systematic review. *Applications of Ergonomics*, 97. https://doi.org/10.1016/j.apergo.2021.103528.

Kim, H., Kim, D., & Lee, J. (2019). 3D printing of synthetic diamonds for industrial applications. *Journal of Industrial and Engineering Chemistry*, 74, 217–222.

Klenam, D., Bamisaye, O. S., Williams, I. van der Merwe, J. W., & Bodunrin, M. (2022). Global perspective and African outlook on additive manufacturing research – An overview. *Manufacturing Review*, 9(35), pp. 1–37. http://doi.org/10.1051/mfreview/2022033.

Kudryavtseva, E., Popov, V., Muller-Kamskii, G., Zakurinova, E., & Kovalev, V. (2020). Advantages of 3D printing for gynecology and obstetrics: A brief review of applications, technologies, and prospects. *Proceedings of the 2020 IEEE 10th International Conference*, pp. 9–13, Nanomaterials Applications and Properties, Sumy, Ukraine.

Mngomezulu, T. (2018). Ubuntu and product design: A cultural perspective. *Journal of Design and Culture*, 10(4), pp. 365–378.

Ndlovu, P., & Mkhize, K. (2020). The role of Ubuntu in sustainable product design. *Journal of Sustainable Design*, 12(2), pp. 45–54.

Nkosi, N., & Mkhize, S. (2015). The spirit of Ubuntu in product design: A study of South African design practices. *Journal of African Design Research*, 5(2), pp. 32–41.

Popov, V. V., Kudryavtseva, E. V., Katiyar, N. K., Shishkin, A., Stepanov, S. I., & Goel, S. (2022). Industry 4.0 and digitalisation in healthcare. *Materials*, 15, pp. 1–21.

Sakin, M., & Kiroglu, Y. C. (2017). 3D printing of buildings: Construction of the sustainable houses of the future by BIM. *Energy Procedia*, 134, pp. 702–711. https://doi.org/10.1016/j.egypro.2017.09.562.

Stavropoulos, P., & Foteinopoulos, P. (2018). Modelling of additive manufacturing processes: A review and classification. *Manufacturing Review*, 5, pp. 1–26.

Thomas, M. J., & Srinivasan, S. (2018). Synthetic diamond production using additive manufacturing techniques. *Diamond and Related Materials*, 82, pp. 50–56.

Vayre, B., Vignat, F., & Villeneuve, F. (2012). Designing for additive manufacturing. *Procedia CIRP*, 3, pp. 632–637.

Wang, J. C., Dommati, H., & Cheng, J. (2019). A Turnkey manufacturing solution for customised insoles using material extrusion process. In *3D Printing and Additive Manufacturing Technologies*, pp. 203–216. Singapore: Springer.

Wang, L., Wang, D., & Li, Y. (2022). Single-atom catalysis for carbon neutrality. *Carbon Energy*. 4:1021–1079. https://doi.org/10.1002/cey2.194.

Wang, Y., Ahmed, A., Azam, A., Bing, D., Shan, Z., Zhang, Z., Tariq, M. K., Sultana, J., Mushtaq, R. T., Mehboob, A., Xiaohu, C., & Rehman, M. (2021). Applications of additive manufacturing (AM) in sustainable energy generation and battle against COVID-19 pandemic: The knowledge evolution of 3D printing. *Journal of Manufacturing Systems*, 60, pp. 709–733. https://doi.org/10.1016/j.jmsy.2021.07.023.

World Economic Forum. (2019). *Global Competitiveness Report 2019*. https://www3.weforum.org/docs/WEF_TheGlobalCompetitivenessReport2019.pdf.

Wu, B., Myant, C., & Weider, S. (2017). *The Value of Additive Manufacturing: Future Opportunities*. Imperial College London, Briefing Paper, (2). https://spiral.imperial.ac.uk/bitstream/10044/1/53611/2/IMSE-AMN%20The%20value%20of%20additive%20manufacturing-future%20opportunities.pdf.

WWF. (2023). *Enhancing Energy Access Through Affordable Renewables to Reduce Emissions and Benefit People and Nature*. https://wwf.panda.org/discover/our_focus/climate_and_energy_practice/what_we_do/changing_energy_use/energy_access_africa/.

Zhang, Y., Wu, X., & Li, Z. (2020). Additive manufacturing of synthetic diamonds: A review. *Journal of Materials Science & Technology*, 36(6), pp. 1051–1058.

Zhang, Z., & Chen, Y. (2016). Additive manufacturing of ceramics: A review. *Journal of the American Ceramic Society*, 99(3), pp. 905–921.

Zungu, T., & Mkhize, L. (2016). Incorporating Ubuntu values into product design: A case study of a South African furniture company. *Journal of Design Management and Professional Practice*, 8(3), pp. 123–132.

PART 5

Taking the Product to the Market

12

BUILDING CORPORATE BRAND REPUTATION IN EMERGING ECONOMIES

Debra Diana Ralitsha, Odireleng Marope and Richie Moalosi

Introduction

Successful brands exist because entities have made a deliberate effort to design strategies to make them visible and relevant to their customers and to penetrate the competitive global market. Emerging economies, especially in Afrika, are dominated by businesses operated by small micro enterprises (SMEs). Usually, such businesses treat branding and package design as an afterthought. As a result, they experience no or little business growth. SMEs need to learn more about branding and to package their products. Afrikan design professionals have a critical role in guiding SMEs in emerging economies to transit from running businesses informally to a professional setup and consider branding and package design at the early stages of product development.

Against this background, the discourse in this chapter examines how designers can tap into Afrika's Indigenous knowledge systems (IKS) to develop brands and packaging designs. Several case studies from Botswana and Ghana will be reviewed on branding Afrikan indigenous products. Afrikan designers are encouraged to learn from Afrikan philosophies such as Ubuntu, Afrikology and Afrikan ethos to develop brands that are sensitive to context and can also appeal globally. A culture-oriented design model has been proposed, which can assist designers in developing cherishable brands and package designs.

Brand Development and Indigenous Knowledge Systems

Design experimentation for branding in emerging economies will require integrating local knowledge to inform the design process. IKS within the Afrikan culture permit the inclusion of local philosophies, which makes the brand unique.

DOI: 10.4324/9781003270249-18

In many instances, these local knowledge systems can present a different perspective to the design while simultaneously preserving and promoting local knowledge on the global market. In design experimentation, we often make the "familiar = unfamiliar" and the "unfamiliar = familiar". One way to achieve these different dimensions is through co-creation, where knowledge is co-constructed with users to achieve the desired value prepositions (Debrah et al., 2017). These design processes and concepts permit the designer to externalise their ideas in ways which can be an abstraction of the original idea inspired by IKS in the local context. These design processes lead to selecting a brand/product concept and ultimately give the brand an Afrikan "SOUL". The "soul" inspired by authentic IKS will add value to the finished brand/product (Figure 12.3).

Further, the brand experimentation process informed by IKS can be inspired by selected symbols, ideologies, proverbs and values in a particular Afrikan culture. Drawing inspiration from other sources provides a new perspective on the brand/product. For instance, Afrika is rich in cultural symbols, such as the Adinkra, Nsibi and Bantu symbols, which can inspire brand design and development (Mafundikwa, 2004). Branding locally manufactured products informed by IKS can add value to the product, making it unique on the international market. The value addition due to efficient branding will enable locally manufactured brands to compete in the international market.

Branding Indigenous Afrikan Products

This section examines the branding of local products/packaging designs from Afrika. Afrikan SMEs have been developing several products and services. However, these products cannot compete internationally due to inadequate branding and packaging design. Affective branding is critical to promoting goods and services internationally and increasing revenue for packaging manufacturers. On the other hand, it will provide sustainable jobs for SMEs in emerging economies.

Case Studies: Branding and Packaging in Emerging Economies

The authors worked with a group of designers who "experimented with designing" and "re-designing" local brands/products manufactured locally for sale to promote export. The design prototypes were co-created with the manufacturers and potential users to develop the desired brand or packaging suitable for their needs (Debrah et al., 2015). A few case examples, which are outcomes of the design experimentations, have been discussed in this chapter. These case studies comprise corporate branding of food and fashion products which needed packaging for export. The authors will focus on the brand and packaging design procedures used by Botswana and Ghana designers. Similar processes were applied to the rest of the reviewed case studies for the brand and packaging development of the local Afrikan products.

Case Study 1: SHEROO Project Description

Brand Name/Product: SHEROO

A case study was undertaken with a Botswana SME selling an organic drink. This was meant to demonstrate the importance of branding and package design on how they can transform a business entity. Shepherd Beverages is a micro-enterprise producing an organic beverage which tastes like coffee. The beverage is produced by roasting the roots of a shepherd tree and then grounding them into a powder form. The product is believed to contain nutritional and medicinal values. The enterprise faced the challenge of penetrating the market by selling powdered products through chain supermarkets. At first, the chain supermarkets rejected the product due to poor package design and branding. The University of Botswana's Department of Industrial Design and Technology has a module for branding and package design. Each year, an SME is identified which can benefit from students' work in this module. This is part of the community service, or corporate social responsibility, the university is doing to give back to the community. Involving students in such

FIGURE 12.1 SHEROO re-designed product package

Source: Unangoni Thibatsela

a product educates future designers to be compassionate and assist disadvantaged entities or communities.

The product was initially packaged in bottles and used a sticker for branding. There was much interest in the product, but it could have performed better in the market with more attractive branding and package design. The product brand name was Shepherd Beverages. The product market performance improved after the involvement of design students. The brand was renamed SHEROO and was developed from the words *SHEpherd tree* and *ROOts*. The product identity in the form of a logo was also developed. The package was designed taking into consideration the natural aspect of the product. The designers considered the targeted market and trending related products, which informed the design outcome. After improving the package design and branding, the local chain supermarkets started buying the product.

Case Study 2: "Give Your Brand a Soul" Project Description

A young professional designer and director of Blue Jacket Designs in Botswana found it fit to teach the public about branding online as part of his corporate social responsibility. When the designer deals with clients, he follows these four distinct phases.

Meeting With the Client

The first phase is an in-person meeting with the client to get a brief on the work. This assists in deciding if the client's work to be done is aligned with the designer's design philosophy.

Diagnosis

In this phase, the designer finds out the real cause of the problem that leads the client to seek his services. They usually come with a self-diagnosis and solution for the designer to work on. However, the designer has grown to know that the client's diagnosis and solutions may need to be more balanced and address the problem. The designer does diagnostics to ensure that the problem identified is worth solving and that the solution will be appropriate and sustainable.

Synthesis

In this phase, the designer synthesises all the data gathered in the diagnosis phase into insights that can co-create a design brief that outlines the strategy used to solve the problem. The strategy is then shared with the client for appreciation and approval.

Solution Creation

After the design brief is approved, the designer moves on to co-create a solution based on the results of the synthesis phase. Depending on the project, the solution could be a logo, brand identity, creative strategy, etc. presented to the client. The solution is objectively co-created to deliver a solution with consumers' or clients' interests. This process removes all client or designer biases in creating the solution. The co-creation processes and outcomes obtained from working with clients inform the brand and product development, as illustrated in Figures 12.2 and 12.3).

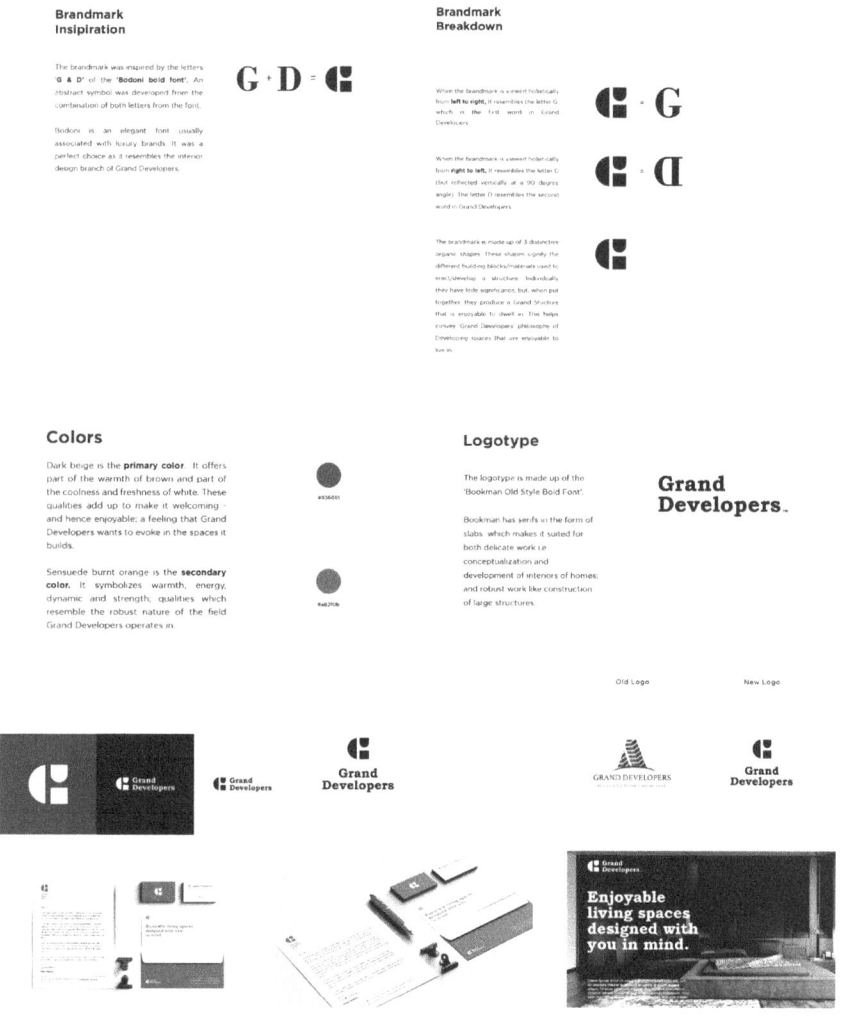

FIGURE 12.2   Grand developers brand – work achieved through co-creation conviction

Source: Jonathan Mabhekede

Design
Brief

To design **Adepa Innovate+** brand identity to carry
its motive to help the youth DREAM, MAKE,
INSPIRE and work together

Brandmark
Breakdown

The brandmark is an abstract representation of 4
people with their arms outstretched, taken from
the **adinkra symbol "ATI KORO NKO AGYINA"**

Brandmark
Inspiration

The brandmark is inspire my the
Adinkra symbol "ATI KORO NKO
AGYINA" also by symbols of people
overlapping arms and heads

The brandmark is a modernised inverted version of
the Adinkra symbol with rectangles used to
represent the limbs and head of people
represented instead of ovals and circles.

The brandmark is also represented by 3 shapes
together which form a creative layout of balance
and Stability.

Colors

Blue the **primary color,** signifies peace
harmony and togetherness, it also brings a
cool feel to our brand. It is also one of the
most prominent colors in the beauty of
nature and how inspiring it can be.

#62ADEC

Dark shade of gray and white are the
secondary colors, these two present the
extremes of the color palate signifying how
vast our creativity can be.

#273426   #FFFFFF

Logotype

The logotype is made of Nexa Heavy and
Nexa Bold.

Adepa Innovate +

Nexa is a San Serif type font, bold and
stable is it unique and gives a sense of
dynamism. The Nexa family has a wide
range of fonts from thin to black.

FIGURE 12.3 Adepa Innovate + achieved through co-creation and inspired by Afrikan IKS

Source: Author's image (Brand and Packaging Design Unit, KNUST: 2022)

Besides the design phases adopted by the designer, which serve as a guide to other designers, he openly shares his approaches through social media networks. By doing so, other designers benefit from the designer's posts and, at the same time, promote professional interaction. The action embraces the Afrikan culture of Ubuntu, which is not a concept easily distilled into a methodological procedure, but rather a bedrock of a specific lifestyle or culture that seeks to honour human relationships as primary in any social, communal or corporate activity (Nussbaum, 2003).

Case Study 3: Adepa Innovate + Project Description

Brand Name/Product: Adepa Innovate +

Adepa Innovate + is a future re-branded design and innovation centre to train and equip young people and designers with the 21st-century soft skills required for the workforce in Ghana. The design and innovation centre aims to create innovative solutions and prepare young professionals to design for society's common good ("Adepa") in emerging economies. The brand is inspired by an Adinkra symbol – "*Eti koro eko Agyina*", which can also be interpreted as "working together" for the "common good of society." The Adinkra symbols can be found in Ghana, in West Africa, and are part of everyday activities and woven into the social-cultural fabric of the country (Figure 12.3).

Case Study 4: Sorgvirg Pito Project Description

Brand Name/Product: Sorgvirg Pito

Pito is a food drink made from cereal in Ghana. It is a traditional alcoholic beverage product brewed locally but can be found among the populace in the northern region of Ghana. *Pito* is prepared from carbohydrate-rich cereal crops such as millet, guinea corn or maisemaise. The drink serves as an energy drink, which has gained popularity among young people in their region. *Pito* contains healthy nutritional values. Regular intake could help improve the health status of consumers (Adazabra et al., 2014). In this case study, the *Pito* brand was designed using the iterative design process to create the various elements that constitute the package. The subsequent headings have briefly described these processes (Figure 12.4).

1 **Design process/brand name**: the design process began with identifying a suitable name to make it easier to design the brand. The name "Sorgvirg", derived from the words **sorg**hum and **virg**in, was was ultimately chosen as

the brand name. The indigenous name of the product, "*Pito*", which means made from sorghum seeds and virgin, was then added to the brand name. The proposed brand name denotes a sense of newness and an organic product. The name was further iterated and drew inspiration from local elements, such as a calabash, to inform the final brand (Figure 12.4).

2 ***Label design***: these were created through the iterative design process until the desired sample was obtained. A transparent glass bottle was chosen for

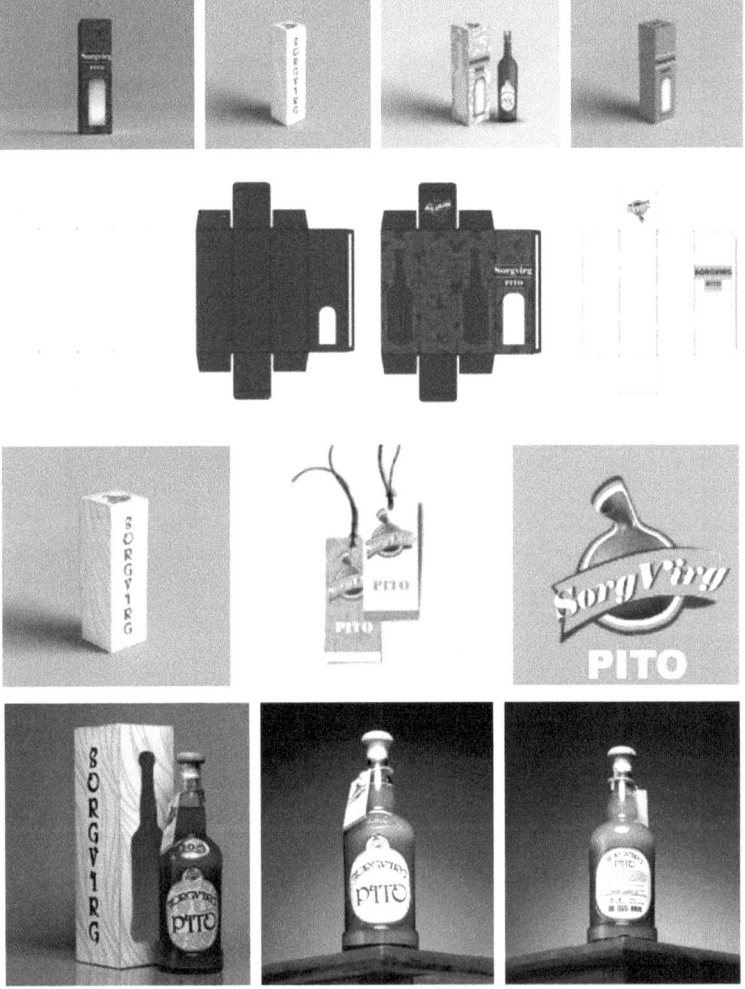

FIGURE 12.4   The Pito brand's final packaging design and components (https://bit. ly/3vTdxPO)

Source: Author's image (Brand and Packaging Design Unit, KNUST: 2022)

the drink because it could be recycled and does not leak chemicals like plastic packaging. The labels were designed and printed on transparent material so that the *Pito* in the glass bottle would be visible to consumers.

3 **Tag label design**: samples were created, and the suitable one was affixed to the bottle. The bottle's cork is covered with paper to protect it from spills. The tag was firmly fixed onto the bottle and around the cork using a local thread.

4 **Secondary package design** : the secondary package was primarily designed to carry the *Pito* in the glass bottle. The package was designed as a thick mini box with the sides cut out and covered with transparent plastic. This way the *Pito* was still visible in the secondary package to attract consumers.

5 **Design and development of the prototype** : designs were created on a computer, then printed and cut out to form the prototypes. These were then folded by hand to form the box, which was used as the secondary package to make the product attractive and appealing to the local and export market (Figure 12.4).

Case Study 5: Adinkra Delight Project Description

Brand Name/Product: Adinkra Delight

The Adinkra delight package is an innovative idea to increase the market performance of an already famous product, Adinkra Pie, and other pastries in Ghana. Although a fresh, quality food product by itself, concerns have been raised about the dull package design. The lack of an effective colourful design negatively affects some consumers' purchasing behaviour, leading to low sales of the product. A new package was designed with a sleek ethnocentric surface inspired by the Adinkra symbols. Additionally, a second package was developed using the aluminium foil packaging often preferred by the middle class for lunch. It is envisioned that the proposed new package design could help boost sales for the Adinkra delight product (Figure 12.5).

FIGURE 12.5   Re-designed package – Adinkra delight

Source: Adam Rahman's image (Brand and Packaging Design Unit, KNUST: 2022)

Case Study 6: Prekese Project Description

Brand Name/Product: Prekese

Using various herbs for nutritional and medicinal purposes has been a cultural practice in Ghana and many Afrikan countries for centuries. The product Prekese is botanically known as Tetrapleura tetraptera. However, it is popularly referred to as "Prekese" in the local Ghanaian dialect. Prekese is said to have many nutritional values and has become a household product for many Ghanaians and across West Afrika. Regardless of these benefits, the indigenous product Prekese is largely un-packaged, un-labelled and sold in the open market. This makes the product susceptible to insects and decay. Branding Prekese as a packaged product will preserve the content and increase the shelf life, making export and promotion possible in an emerging economy (Figure 12.6).

Case Study 7: Abeduro Project Description

Brand Name/Product: Abeduro

The product Abeduro is also known as turkey berries. Turkey berries can be found in many parts of West Afrika and other countries on the continent. The scientific name for turkey berries is *Solanum torvum*, which is popularly known as "Abeduro" or "Kwahu Nsosoa" in Ghana. Turkey berries ("Abeduro") have

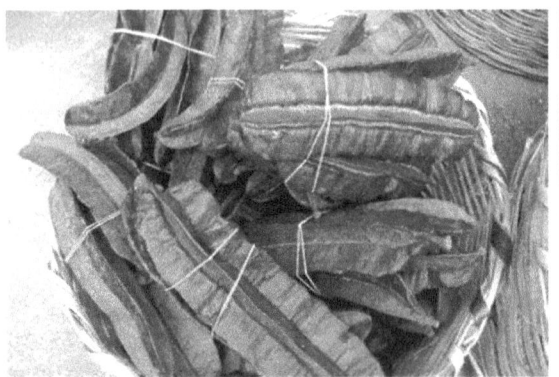

FIGURE 12.6   Newly designed package – Prekese

Source: Adam Rahman's image (Brand and Packaging Design Unit, KNUST: 2022)

FIGURE 12.7   Newly designed package – Abeduro (Turkey berries)

Source: Adam Rahman's image (Brand and Packaging Design Unit, KNUST: 2022)

high levels of minerals and nutritional value and are a household product in Ghana. This product is often sold on the open market and has no packaging to protect and increase the shelf life. Introducing a package for the product is a creative way to preserve and promote the product and boost sales. The proposed package was designed for the juice version of Abeduro. Projections are that the newly designed package will increase sales performance in the export market (Figure 12.7).

Case Study 8: Sedorh Ahenema Project Description

Brand Name/Product: Sedorh Ahenema

Sedorh Ahenema is a beautiful traditional pair of slippers worn by virtually everyone for occasions such as festivals, funerals, marriage ceremonies, church and other ceremonies in Ghana and some parts of West Afrika. The slipper is made of leather, and the soles are carved into the shape of a figure 8, which represents stability. The slipper, which is locally manufactured, is often sold without a brand name and has no package. Therefore, the product is desired and can be exported with the appropriate packaging design. The newly designed product was achieved through the creative iteration of local cultural elements, which inspired the brand/package design. The proposed Sedorh Ahenema package can be exported to promote this locally manufactured product from an emerging economy (Figure 12.8).

FIGURE 12.8   Newly designed package – Sedorh Ahenema

Source: Adam Rahman's image (Brand and Packaging Design Unit, KNUST: 2022)

Afrikan Philosophies and Brand Development

1 **Afrikology in brand development**: The term refers to an epistemology of knowledge generation and application firmly rooted in Afrikan ideologies (Nabudere, n.d., p. 1). Dani Wadada Nabudere, a Ugandan scholar who supports the ideology of Afrocentrism, formulated the concept of Afrikology. Nabudere's worldview of Afrikology can be traced to historical and scientific perspectives of Afrika and its relationship to socio-cultural paradigms. Further, Afrikology philosophy examines the connections between the "cradle of humankind" and the contemporary world. The rationale is to promote healing and re-uniting humankind due to the problematic occurrences in the past caused by social injustices such as forcefulness, mendaciousness and conflict towards embracing indigenous people's origins and developing a psychological sense of wholeness for an Afrikan collective self (Osha, 2018).

2 **Afrikan ethos in brand development**: Design ethics are critical in brand and packaging development. There is a need to establish that the content promised is maintained on the brand or package. Designers are responsible for liaising with manufacturers so that there will be no lapses in the information provided to meet approved ethical standards in design. Designers are encouraged to uphold fundamental ethical principles about the ecosystem of the brand or packaging that they might be working on to maintain the status quo of ethical design aligned with Afrikan world views such as Ubuntu – which is a form of humanism translated as "being self through others", "I am because of who we all are" (Lutz, 2009; Mugumbate & Nyanguru, 2013). Afrikan design ethos inspired by the philosophy

of Ubuntu seeks to create harmony between humanity, the planet and the ecosystems with which it exists (van Niekerk & M'Rithaa, 2009; Debrah, 2021). Aligning brand and packaging design and development to Afrikan ethos will establish the product and bolster sales in its local context while meeting global export standards and trends.

3 **An aspect of Ubuntu in brand development**: Afrika is known for the concept of oneness, Ubuntu, which believes that we exist because of others; everyone is valuable. "Ubuntu is the capacity in Afrikan culture to express compassion, reciprocity, dignity, harmony and humanity to build and maintain community with justice and mutual caring" (Nussbaum, 2003). Lutz (2009) emphasises the notion of Ubuntu by stating that in a true Afrikan community, the individual does not pursue the common good rather than his or her good, but rather pursues his or her good through pursuing the common good. In Botswana, there is an informal setting of coming together and making monthly contributions of money saved and shared at the close of the year. To make a profit, members borrow at a reasonable interest rate. It is called *motshelo*, commonly led and managed by women. It is managed based on trust, with no formal documents such as a constitution or contracts. Big corporations have since noticed this arrangement. They developed packages for *motshelo* groups, such as food or building materials. Banks also created special bank accounts for the same. The concept of *motshelo* continued to attract interest in more formal settings where cooperatives, such as Savings and Credit Co-Operative Societies (SACCOS), a society-oriented arrangement guided by cooperation. For example, cooperative societies embraced the corporate approach by developing brands to aid management.

Brand Development and Culture-Oriented Design Model

1 *Culture-oriented design model*: Theories provide guidance and serve as a research foundation. In line with design research best practices, the culture-oriented design (CoD) model (Moalosi, 2007) is proposed as the framework to guide designers embarking on branding and packaging design-related research inspired by the Afrikan ethos. The CoD model inspired the design and development of some branding case examples described in this chapter. Culture is explored as a tool during the design process, as indicated in the CoD model. Elements of culture are intentionally integrated into designing context and culturally specific solutions. The CoD model has three main principles: socio-cultural factors, integration and cherishable culturally oriented products.

The *socio-cultural factors* represent the user's domain, where the user's needs and expectations are considered elements that may be integrated

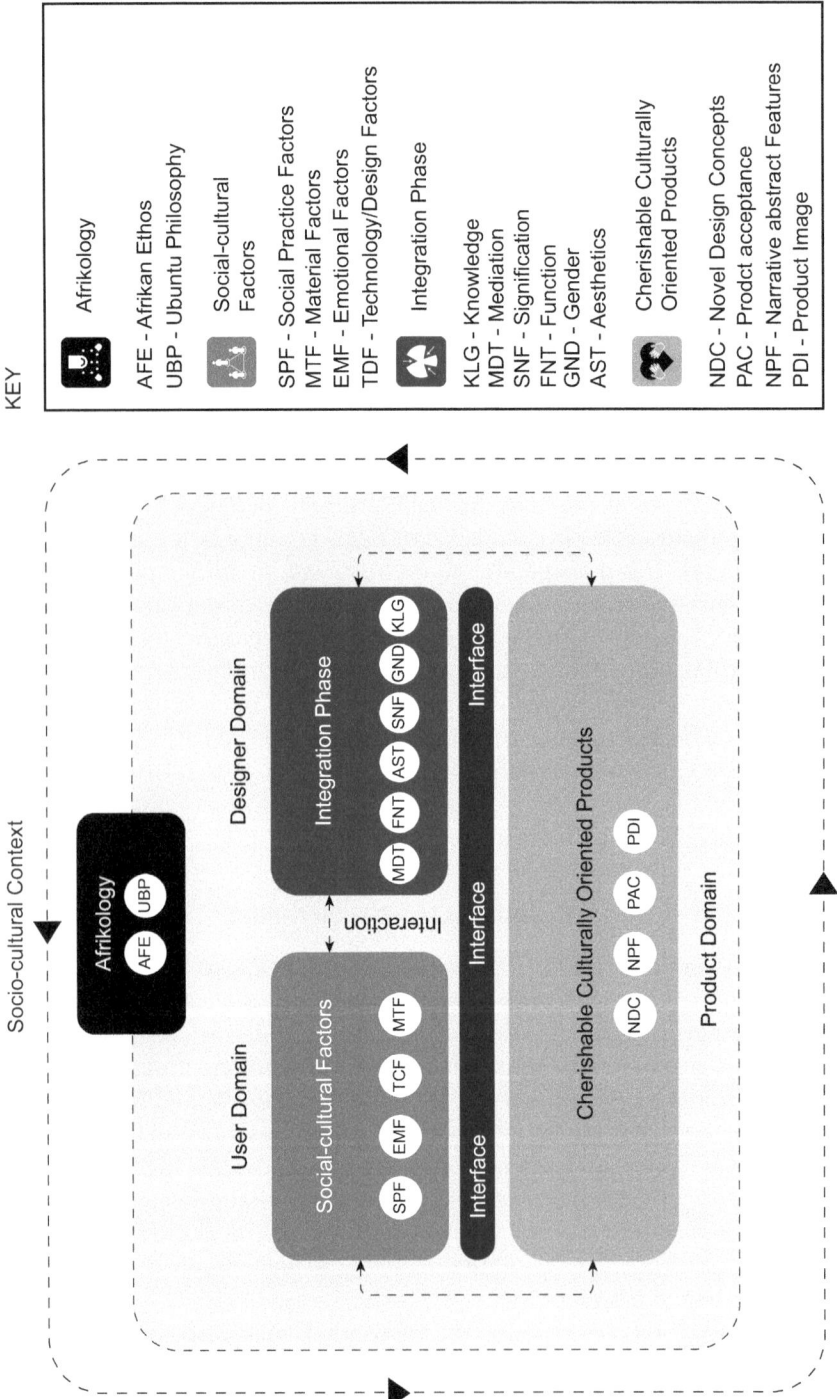

FIGURE 12.9 Afrikology + culture-oriented design model for brand development

Source: Author's image (Author's construct – modeled after Moalosi, 2007, p. 139)

tangibly or intangibly into designing products. This leads to the next phase of the theory, where the *integration process* becomes the designer's domain. The designer then uses the social-cultural factors to reference the user's culture or environment to create culturally oriented designs. The result of the design process becomes a product, considered a *cherishable culturally oriented product*, which then becomes the product domain, thus the third phase of the theory. The product is cherished because it consists of elements with which the user can emotionally connect (Moalosi, 2007).

2 **Afrikology**: Another perspective, which is a proposal and an addition to the model, is the "Afrikology" philosophy in the design context. Afrikology, in this case, examines Afrikan philosophies, which can inform the design process and promote Afrikan culture and values (Rukooko & Komakech, 2017; Osha, 2018). It also includes the integration of Afrikan ethos, which will serve as the foundation of the design and development process for the brand or product development (Debrah, 2021). Additionally, the Ubuntu philosophy comes into play in the design process. The Ubuntu principle in design fosters collaborative activities (co-creation) with users to inform the design outcome for sustainable brand prepositions (Debrah et al., 2015, 2017). Ubuntu, the Afrikan philosophy, advocates for working collectively for the good of humankind and society (Lutz, 2009; M'Rithaa & Jamie, 2017). Designers can explore these concepts as part of the emerging *Afrikology + culture-oriented design model (A+CoD)* model towards designing sustainable brands that are culturally specific and satisfy user needs in emerging economies (Figure 12.9).

Conclusion

Branding and packaging design are necessary to promote and add value to goods and services produced in Afrika. In emerging economies, doing business without the appropriate brand can potentially undermine sales and revenue generation for SMEs. Designers play an essential role in promoting these brands/products/services to make it possible for local products to compete in the international market. The brands ought to be authentic and promote Afrikan culture to make the designs unique and attractive to influence consumer behaviour towards the product. Design can strengthen SMEs and increase sales of their goods and services. Also, affective branding and packaging will positively impact businesses in emerging economies in Afrika. It is envisioned that designers will explore the culture-oriented design model + Afrikology philosophies (A+CoD) to enhance brand and packaging development in Afrika. Finally, standardised branding and packaging policies and collective participation of stakeholders in the spirit of Ubuntu (Lutz, 2009) are required for sustainable brand development by SMEs and prosperity in Afrika's emerging economies.

Reference List

Adazabra, A. N., Ntiforo, A., & Bamford, S. A. (2014). Analysis of essential elements in Pito-a cereal food drink and its brands by the single-comparator method of neutron activation analysis. *Journal of Food Science and Nutrition*, 2(3), pp. 230–235. http://doi.org/10.1002/fsn3.95.

Debrah, R. D. (2021). *Design for Health: Co-designing Health Information Services in the Afrikan Context*. Doctoral dissertation, Cape Town: Cape Peninsula University of Technology.

Debrah, R. D., de la Harpe, R., & M'Rithaa, M. K. (2015). Exploring design strategies to determine the information needs of caregivers. In A. Breytenbach & K. A. Chimela-Jones (Eds.), *7th International DEFSA Conference Proceedings*, pp. 67–77. Johannesburg: Design Educators Forum of South Africa (DEFSA). http://doi.org/978-1-77012-137-9.

Debrah, R. D., de la Harpe, R., & M'Rithaa, M. K. (2017). Design probes and toolkits for healthcare: Identifying information needs in African communities through service design. *The Design Journal*, 20(sup1), pp. S2120–S2134. http://doi.org/10.1080/14606925.2017.1352730.

Lutz, D. W. (2009). African Ubuntu philosophy and global management. *Journal of Business Ethics*, 84, pp. 313–328. http://doi.org/10.1007/s10551-009-0204-z.

Mafundikwa, S. (2004). *Afrikan Alphabets: The Story of Writing in Afrika*, 1st ed. (B. Mark, ed.). West New York: Mark Betty. www.MarkBettyPublisher.com.

Moalosi, R. (2007). *The Impact of Socio-cultural Factors upon Human-centred Design*. Doctoral dissertation, Brisbane: Queensland University of Technology.

M'Rithaa, M. K., & Jamie, A. (2017). Advancing the Afrikan lions' narrative: The quest for a sustainable future for all . . . *24th Annual FIDIC-GAMA Conference African Partnerships for Sustainable Growth*, pp. 1–13, Cape Town: FIDEC-GAMA. www.fidic-gama2017.org.za.

Mugumbate, J., & Nyanguru, A. (2013). Exploring African philosophy: The value of Ubuntu in social work. *African Journal of Social Work*, 3(1), pp. 82–100.

Nabudere, D. W. (n.d.). *Afrikology, Philosophy and Wholeness: An Epistemology Institution*, pp. 1–3. Pretoria. Africa Institute of South Africa. https://muse.jhu.edu/book/16776.

Nussbaum, B. (2003). African culture and Ubuntu, reflections of a South African in America, World Business Academy perspectives. *Rekindling the Human Spirit in Business*, 17(1).

Osha, S. (2018). *Dani Nabudere's Afrikology: A Quest for African Holism*. Dakar: Council for the Development of Social Science Research in Africa (CODESRIA). www.codesria.org.

Rukooko, A. B., & Komakech, D. (2017). Symbolic Universes of Hellenism and Afrikology: Metonyms and search for global epistemological field building. *Journal of Humanities and Social Science (IOSR-JHSS)*, 22(8), pp. 61–73. http://doi.org/10.9790/0837-2208066173.

van Niekerk, J., & M'Rithaa, M. K. (2009). The ethical dilemma of a rapidly receding watering hole: Implications for design education. In A. Breytenbach & A. J. Munro (Eds.), *12th National Design Education Forum Conference Proceedings*, pp. 149–157. Graaff Reinet, Eastern Cape: Design Educators Forum of South Africa (DEFSA). www.defsa.org.za/sites/default/files/downloads/2009conference/DEFSA Conference Proceedings 2009.pdf.

13

INTELLECTUAL PROPERTY, TRADITIONAL KNOWLEDGE, TRADITIONAL CULTURAL EXPRESSIONS, AND THE PHILOSOPHY OF UBUNTU

Chinandu Mwendapole, Tebo Motlhaping, Thatayaone Mosepedi, and Tawanda Gombiro

Introduction

It is necessary to create an environment where the generators of knowledge and innovations can receive financial benefits from exchanging and transferring their intellectual property (IP) to ensure investment in creative and innovative knowledge. That means granting protection over their IP for a certain period, in which they can fully exploit their works for returns. Within that context, IP and the resulting intellectual property rights (IPRs) are viewed strictly as private and individual protection and benefit. Yassine (2018, p. 80) observes that "intellectual property rights systems create individual property rights and are different to traditional knowledge that is usually developed and transmitted through communities."

Traditional knowledge is knowledge that pertains

> to a whole tradition and way of life, on how to interact with the natural world, that does not arise from one individual or a single creator but rather over decades, centuries and or millennia through the interaction of individuals and groups.
>
> *(Milius, 2009, p. 193)*

Ibe and Obianyo (2021, p. 131) argue that traditional cultural expressions (TCEs) cover "elements of the traditional and artistic heritage developed and maintained by the community, or by individuals reflecting the traditional artistic expressions of such a community."

Chigangaidze et al. (2022, p. 322) write, "Ubuntu is a philosophy that emphasises the relationality between the individual and the community." Ubuntu's axiom, *umuntu ngubuntu ngabantu*, means that "a person is a

DOI: 10.4324/9781003270249-19

person through other people." The core of the philosophy of Ubuntu includes the desire to ensure harmony between the individual and the community. According to Mabavurira (2020, p. 75):

Ubuntu ethics:

- dwell on the good of the majority (community) over personal good (community good);
- ensure the course of action chosen should treat individuals equally (social justice/fairness);
- respect, love, care, and compassion for others, especially the vulnerable (respect); and
- should bring no least harm to all the parties involved (no least harm).

Knowledge

Nonaka et al. (1996, p. 205) note that "there exist two types of knowledge, tacit knowledge (intuitions, unarticulated mental models, or embodied technical skills) and explicit knowledge (i.e., a meaningful set of information articulated in clear language including numbers or diagrams)." There are four methods of knowledge conversion (Nonaka et al., 1996) (Figure 13.1).

Dlamini (2017, p. 83) suggests that "these four elements of knowledge creation are fundamental to exchanging knowledge, not only within organisations but also the affairs of life." Tacit and explicit knowledge are critical

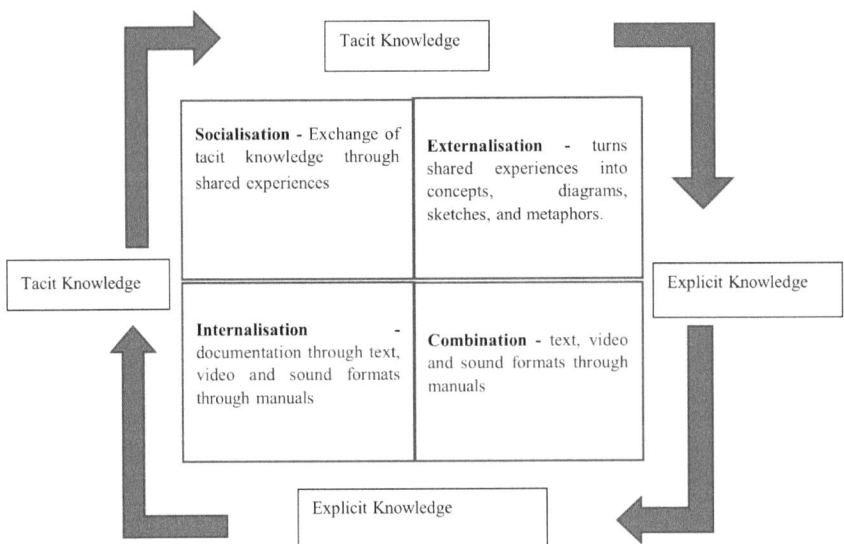

FIGURE 13.1 Four nodes of knowledge conversion

Source: Adapted from Nonaka et al., (1996, p. 207)

drivers of economic development; however, economists describe knowledge as a non-rival good; in other words, once it is made explicit or enters the public domain, its use cannot be limited, like physical products (Withers, 2006).

Intellectual Property and Knowledge

For tacit knowledge to be considered to have an economic value, it is allocated intellectual property rights once made explicit (World Intellectual Property Organization [WIPO], 2020a). Examples of explicit knowledge that can be considered intellectual property include aesthetic and functional intellectual knowledge that comprises artwork, music, literary works, designs, symbols, and scientific and technical knowledge (Christie, 2011, p. 8). The primary laws that guard and administer the rights of the authors and inventors include copyright and industrial property laws.

Christie (2011, p. 7) argues that "intellectual property is like physical property, in that exclusive rights may be granted, and the owner may transfer those rights to other persons. However, intellectual property is dissimilar to physical property in three important ways: non-rivalrous, inexhaustible, and non-excludable." In other words, IP can be used by more people at the same time as music, it has no expiry date, and one cannot limit its access, which makes it vulnerable to exploitation by third parties. For knowledge to have an economic value, its transfer and exchange are, therefore, regulated by what Withers (2006, p. 6) identifies as "state-granted, time-limited monopolies for information goods in the form of intellectual property rights." These monopolies granted by the state represent an individualistic concept of creativity. For this reason, Ncube (2017, p. 263) observed that from an Ubuntu viewpoint, "intellectual property laws and principles have failed to gain traction in Africa because they are based upon western ideas of an individual's rights to intangible property that fail to factor in widespread communitarian perspectives."

Traditional Knowledge/Traditional Cultural Expressions

Singh and Singh (2020, p. 71) argue that "traditional knowledge is an essential source of information that has been developed from various generations of the local communities and indigenous people in many parts of the world. These people have secured and protected traditional knowledge for thousands of years." According to the World Intellectual Property Organisation (WIPO) (2020b, p. 14):

Traditional Knowledge (TK) can be, for example, agricultural, environmental, or medicinal knowledge or knowledge associated with genetic resources. Examples include:

- Knowledge about traditional medicines.
- Traditional hunting or fishing techniques.
- Knowledge about animal migration patterns or water management.

On the other hand, WIPO (2020a, p. 15) views TCEs as

the forms in which traditional culture is communicated. They can be, for example, dances, songs, handicrafts, ceremonies, designs, tales or any other artistic or cultural expression. TCEs are integral to indigenous and local communities' cultural and social identities and heritage, reflecting core values and beliefs. TCEs are handed down from one generation to the other and used, maintained, or advanced by their owners. They are continuously evolving, developing, and being recreated.

WIPO (2020a, p. 13) also notes:

TK is a living body of knowledge that is advanced, sustained, and reflects social identity. In other words, TK is understood as:

- knowledge, know-how, skills, innovations, or practices.
- that are passed between generations.
- in a traditional context.
- and that forms part of the indigenous, traditional lifestyle of the local communities who act as their custodian.

Ibe and Obianyo (2021, pp. 131–132) observe that:

traditional cultural expression systems:

- are passed either orally or by imitation from one generation to the other.
- reflect a community's cultural and social identity.
- consist of attributes of a community's heritage.
- are also made by 'unknown authors' and/or by societies and/or by individuals mutual recognised as having the right, responsibility, or permission to do so; and
- are often not developed for commercial purposes but as channels for religious and cultural expression, and are continuously evolving, developing, and being recreated within the community.

Ubuntu Philosophy

Ubuntu places the community rather than the individual at the centre of knowledge creation and views the production of aesthetic and scientific knowledge as a product of the collective rather than the individual (Mabvurira, 2020, p. 75). Overson (2011, p. 90) notes that

while individualism makes the individual the foundation of all reality in society, Ubuntu defines community as describing an individual. An individual develops to become human jointly with others and not alone. Accordingly, an individual is described by what his/her community is or is not.

In his view, "the commons belong to everyone in the community and not to one individual. Thus, the utilisation and conservation are a communal and collective responsibility." The communitarian nature of Ubuntu implies that people's behaviour must have a clear goal of adding value to society (Coleman, 2021, p. 10). Abubakre et al. (2021, p. 8430) assert that:

Ubuntu represents a perspective in which individuals are as strong or weak as their community and vice-versa. That is, Ubuntu emphasises that one's success depends on the cooperation and contributions of the community. Economic value derives from the community's capacity instead of individuals' abilities.

Ubuntu knowledge management theory promotes the relationship and linkage between people living in the same community, which relies on the belief in harmony with others to ensure individual honesty and a successful way of life (Nansubuga & Munene, 2020, p. 108). "While it is obvious that customary law and Ubuntu are not comparable, it must be equally apparent that, as a fundamental value that informs the regulation of Afrikan dispute resolution and interpersonal relations, ubuntu is inherent to customary law" (Himonga et al., 2013, p. 371). Hailey (2008, pp. 12–16) notes that the

five general areas where Ubuntu has practical application include:

- Helping people value themselves through their relationship with a particular community and compassion and respect for others.
- Community building reduces self-interest, helps community transformation, and attains a degree of community connection.
- Encouraging collective work and consensus building through sharing ideas between community members.
- Potential conflict mediation and reconciliation through consensus-based mediation based on rather than confrontational litigious processes.
- Organisational effectiveness and productivity that helps promote effective team working, the transfer of information, the adoption of new ideas and inventive ways of resolving conflict.

Conventional and Non-conventional Protection Tools of TK and TCEs and the Philosophy of Ubuntu

Communities wishing to protect their TKs and TCEs based on an Ubuntu standpoint have several strategies they can apply. The first strategy, the 'offensive or protective strategy', uses conventional IP protection tools based on Western principles. The second strategy, called the 'defensive strategy', uses non-conventional protection tools that do not fall under conventional IP protection tools.

The main aim of conventional IP protection is to confer exclusive rights to the community that allow them to issue licenses to third parties and control unsanctioned, unwelcome, and unprofessional use by unlicensed holders or free renters (WIPO, 2020a). In contrast, non-conventional protection tools impede external third parties from the right to possess or keep community TK illegally and TCEs, when it would be to the detriment of the community (WIPO, 2020a; Ibe & Obianyo, 2021).

Offensive Strategy Using Conventional Protection Tools

The primary function of the offensive strategy is to protect access to TK and TCEs by using conventional IP such as copyrights, patents, trademarks, industrial design rights, geographical indicators, and trade secrets (Table 13.1).

TABLE 13.1 Conventional IP protection tools covering TK and TCEs

Conventional IP Protection Tools	Duration	Examples
Copyrights are automatic rights that protect the economic and moral rights of the author. The rights are automatically granted when the work is recorded in some explicit format (Ibe & Obianyo, 2021).	Life of the creator + 50 years after their death TK or TCEs whose copyright has expired can be issued new copyright subject to state approval.	Music, dramatic works, audio and visual recordings, photographs, drawings, paintings, signs, sounds, databases, and other folklore-related knowledge
Registered industrial design rights protect the ornamental or aesthetic appeal of an object. The fundamental question for obtaining a design right is whether the appearance is unique	Require registration and last for ten years	Basket patterns, batik designs, traditional utensils, leather goods, furniture, and clothing
Patents are considered the primary choice for indigenous communities, as they offer exclusive community rights to the novel and innovative knowledge, be it an artefact or a process (Islam & Zaman, 2020).	Require registration and last for 20 years	Traditional skeleton construction technologies, knowledge involving genetic resources such as Hoodia plants, and traditional food preservation techniques

(Continued)

TABLE 13.1 (Continued)

Conventional IP Protection Tools	Duration	Examples
Trademarks are a sign that must be distinctive. According to Islam and Zaman (2020, p. 39), "trademarks may be used to protect symbols or signs of manufactured goods and services offered by local people or indigenous communities."	Require registration and renewable every ten years after the initial formal registration	Rooibos, Chibuku, DSTV, Moringaconnect, Purasoda, MTN
Kariyanna (2021, p. 298) maintains that geographical indications represent "a type of intellectual property rights that do not protect novelty, but cumulated goodwill developed in many years. This goodwill is the result of an acknowledged or perceived link connecting a product and a geographical area."	Require registration and renewable every ten years	Champagne from France, "Mazoe" from Zimbabwe, Mysore silk sari from India, 'Swiss-made' products from Switzerland, such as watches
Trade secrets (TS) refers to any business information that derives its value from being kept a secret from a competitor, such that if this information is available to the competitor it will afford them some commercial advantage (Parr, 2018, p. 34).	No limit; protection goes on as long as the secret is kept	Usage or knowledge of the healing properties of plants, tacit knowledge of traditional healers

Source: Authors' work

Defensive Strategy Using Non-Conventional Protection Tools

The defensive strategy uses non-conventional protection tools such as contractual agreements; databases; customary laws; and national, regional, and international protocols on TK and TCEs (Shaikha & Singhalb, 2019; Bagley, 2019).

- According to Islam and Zaman (2020, p. 47), "contractual agreements are commonly used to enforce benefit-sharing agreements and non-disclosure agreements which act as trade secrets. These contracts clarify and elucidate the points of utilising the knowledge and specifics for sharing benefits. Contract-based agreements can be utilised as an additional apparatus to protect TK."

- Kurnilasari et al. (2018, p. 115) note that a "database is a gathering of related data, specifically traditional knowledge best described as a collection or compilation of information related to traditional knowledge." Databases, however, need to be upgraded regularly and digitally unified. In addition, they require adequate financing and skilled human resources to maintain them (Kurnilasari et al., 2018).
- Muller (2013, pp. 13–16) observes: "the registration of the traditional knowledge with national bodies in charge of maintaining databases for traditional knowledge – intellectual property related issues to ensure defensive protection against any misappropriation or misuse of patentable knowledge. It, however, does not offer exclusive rights, control, and compensation."
- "Customary laws, practices and protocols define how traditional communities advance, hold and transmit TK and TCEs" (WIPO, 2020b, p. 30). Regional protocols that cover TK and TCEs include the African Regional Intellectual Property Organization (ARIPO) Swakopmund Protocol on the Protection of Traditional Knowledge and Expressions of Folklore (2010), which regulates the transfer and exchange of TK and TCEs to third parties in member states.
- International protocols covering TK include the Nagoya Protocol on the Use of Genetic Resources, which according to Seifert et al. (2020, p. 1), "is a legal framework that governs the access to genetic resources, and a fair and impartial sharing of benefits arising from their utilisation."
- Muller (2013, pp. 13–16) urges "the use of community protocols or control and organisational tools detailing how specific communities want their knowledge to be used in community projects. The community protocols do not cover the other co-creators but provide a legal framework for community projects and activities."
- Customary laws can play an important role in conjunction with existing IPRs as they consider the essence of TK and TCES and the value systems of the TK holders (Kariuki, 2020).

There are many reasons why conventional IP systems and TK are incompatible. IP laws, such as copyright, protect an idea's expression rather than the idea itself. IP law protects the rights of known individuals. IP rights are limited and may be assigned or sold to other people. On the contrary, TK and TCEs are communally owned by communities and are an essential part of cultural identity.

(Sali & Filipo, 2020, p. 562)

Gogoi, M., & Kaushik, I. (2021). Protection of traditional knowledge in the State of assam and role of geographical indication. *International Journal of Recent Advances in Multidisciplinary Topics*, 2(4), pp. 35–41.

Hailey, J. (2008). *Ubuntu: A Literature Review*. Document. London: Tutu Foundation.

Himonga, C., Taylor, M., & Pope, A. (2013). Reflections on judicial views of Ubuntu. *Potchefstroom Electronic Law Journal*, 16(5), pp. 369–427.

Ibe, I. U., & Obianyo, C. I. (2021). Traditional knowledge, traditional cultural expression, and intellectual property in Nigeria. *De Juriscope Law Journal*, 1(2), pp. 124–140.

Islam, M. T., & Zaman, M. (2020). Protection of traditional knowledge: Finding an appropriate legal framework for Bangladesh. *Dhaka University Law*, 31, pp. 29–48.

Kariuki, F. (2020). *Protecting Traditional Knowledge in Kenya: Traditional Justice Systems as Appropriate Sui Generis Systems*, pp. 91–105. https://www.wto.org/english/tratop_e/trips_e/colloquium_papers_e/2019/chapter_7_2019_e.pdf.

Kariyanna, K. S. (2021). A study on geographical indication as a tool of protection of traditional knowledge. *IJIRT*, 8(5), pp. 298–303.

Kawooya, D. (2013). Ethical implications of intellectual property in Africa. *Information Ethics in Africa: Cross-cutting Themes*, pp. 43–57.

Kurnilasari, D. T., Yahanan, A., & Rahim, R. A. (2018). Indonesia's traditional knowledge documentation in intellectual property rights perspective. *Sriwijaya Law Review*, 2(1), pp. 110–130.

Mabvurira, V. (2020). Hunhu/Ubuntu philosophy as a guide for ethical decision-making in Social Work. *African Journal of Social Work*, 10(1), pp. 73–77.

Milius, D. (2009). Justifying intellectual property in traditional knowledge. *Intellectual Property Institute*, pp. 185–216.

Muller, M. R. (2013). Protecting shared and widely distributed traditional knowledge: Issues, challenges, and options. *International Centre for Trade and Sustainable Development*, pp. 1–28.

Nansubuga, F., & Munene, J. C. (2020). Awakening the Ubuntu episteme to embrace knowledge management in Africa. *Journal of Knowledge Management*, 24(1), pp. 105–119.

Ncube, C. B. (2017). Calibrating copyright for creators and consumers: Promoting distributive justice and Ubuntu. *What If We Could Reimagine Copyright*, pp. 253–280.

Nonaka, I., Umemoto, K., & Senoo, D. (1996). From information processing to knowledge creation: A paradigm shift in business management. *Technology in Society*, 18(2), pp. 203–218.

Overson, S. (2011). Commons thinking, ecological intelligence and the ethical and moral framework of Ubuntu: An imperative for sustainable development. *Journal of Media and Communication Studies*, 3(3), pp. 84–96.

Parr, R. L. (2018). *Intellectual Property: Valuation, Exploitation, and Infringement Damages*. New Jersey: John Wiley & Sons, Inc.

Sali, S., & Filipo, P. (2020). Protecting traditional knowledge: An analysis of the pacific regional framework for the protection of traditional knowledge and expressions of culture. *Victoria University of Wellington Law Review*, 51, p. 559.

Seifert, H., Weber, M., Glöckner, F. O., & Kostadinov, I. (2020). An open-source GIS-enabled lookup service for Nagoya protocol party information. *Database, Journal of Database and Curation*. Jan 1;2020:baaa014. http://doi.org/10.1093/database/baaa014.

Shaikha, S. A., & Singhalb, T. K. (2019). Study on the various intellectual property management strategies used and implemented by ICT firms for business intelligence. *Journal of Intelligence Studies in Business*, 9(2), pp. 30–42.

Singh, B., & Singh, N. P. (2020). *Contemporary Challenges to Traditional Knowledge in India-a Legal Insight*. http://dr.ddn.upes.ac.in:8080/jspui/bitstream/123456789/3095/1/article%20-%20UGC%20care.

Udo, E. J. (2020). The dialogic dimension of Ubuntu for global peacebuilding. In *Handbook of Research on the Impact of Culture in Conflict Prevention and Peacebuilding*, pp. 302–322. Pennsylvania: IGI Global.

WIPO. (2020a). *Intellectual Property and Genetic Resources, Traditional Knowledge, and Traditional Cultural Expressions*. World Intellectual Property Organization. www.wipo.int/about-ip/en/.

WIPO. (2020b). *What is Intellectual Property*? World Intellectual Property Organization. www.wipo.int/publications/en/details.jsp?id=4504.

Withers, K. (2006). *Intellectual Property and the Knowledge Economy*. London: Institute for Public Policy Research.

Yassine, J. (2018). IP rights and indigenous rights: Between commercialization and humanization of traditional knowledge. *San Diego International Law Journal*, 20, pp. 71–92.

14

THE BUSINESS CASE FOR DESIGN IN AFRIKA

Keiphe N. Setlhatlhanyo, Polokano
Sekonopo and Sophia N. Njeru

Introduction

This **chapter is** premised on four United Nations (UN) Sustainable Development Goals (SDGs):

- 5: Achieve gender equality and empower/emancipate all women and girls.
- 8: Promote sustainable, inclusive, and sustainable economic growth, full and productive employment, and decent work for all.
- 9: Build resilient infrastructure, promote inclusive and sustainable industrialisation to foster innovation.
- 12: Ensure sustainable production and consumption patterns (UN, 2015).

Globally, medium, small, and micro enterprises (MSMEs) are considered the engine of a country's socio-economic development because they predominate the private sector (Mehta, 2018). Africa's small and medium enterprises (SMEs) account for more than 90% of businesses and contribute about 50% to the gross domestic product (GDP) (Muriithi, 2017). Delavelle and Rouanet (2017) posit that Afrika's women have the potential to unlock and make a massive contribution to the continent's growth and prosperity. Despite its vast potential, Afrika's creative sector's contribution to the "creative economy" has been relatively small (Nzohabonimana, 2016). The sector remains untapped and undocumented.

Entrepreneurship accords design owners/managers several benefits, namely the opportunity of creating one's own creative process (Robi, 2019). MSMEs face myriad barriers in their attempt to grow and remain competitive and profitable: adhere to global standards of excellence; adapt to new

DOI: 10.4324/9781003270249-20

technological standards, quality, and pricing; and change to an innovative product development process (Mehta, 2018). Other conundrums are racial and gender discrimination and personal hardships (Benson, 2017). For example, women entrepreneurs in South Korea have difficulties acquiring financial resources (Brush et al., 2020; Ghosh et al., 2018; Yang et al., 2016). Afrika's design SMEs encounter heavy tax burdens, inconsistent electricity supply, unfriendly business regulations, negative cash flow, a challenging business environment, and limited or unwilling mentors. Fashion designers decry finding disciplined and self-driven employees, particularly for the growth phase of the business (Robi, 2019). The closure of design start-up SMEs could be that design students may sometimes need help to grasp linear entrepreneurship/business units. Thus, graduates cannot apply the limited entrepreneurship knowledge acquired in design start-ups (Smith et al., 2015). Despite the challenges, Angelsmile fashion house in Kenya has operated for over 15 years. The firm's longevity is attributed to building a brand synonymous with superior quality, exceptional design, and consistency for a growing Kenyan market that understands and appreciates local designers (Robi, 2019).

Prudent financial management is critical at all phases of SMEs for their survival. Entrepreneurial activity globally was adversely affected by the COVID-19 pandemic, leading to the mass closure of small businesses in Pakistan and elsewhere (Nasar et al., 2021). However, some entrepreneurs exploited the opportunities: using the lockdown period to create new collections, seeking new clientele, working from home, and embracing ICT.

The entrepreneur's personality provides the structure needed to handle the demands of business ownership (Benson, 2017) in unstructured situations. For instance, an entrepreneur's aversion to or risk appetite. Benson (2017) investigated Black female fashion designers' social and environmental factors that guided their interest in fashion, entrepreneurship, and formal fashion design education. In addition to formal education, a potential fashion practitioner should acquire relevant industry knowledge through a one-year internship in a design house (Robi, 2019).

It is envisioned that this chapter's design entrepreneurs' voices shall offer practical steps and tools to teach, nurture, and inspire design actors: students, educators, up-and-coming and established entrepreneurs, practitioners, and artisans on strategies to launch and successfully navigate Afrika's highly competitive and ever-changing industry. Especially design graduates' paradigm shifts from job seekers to entrepreneurs/job creators.

Business Strategies Adopted by SMEs

Business strategies refer to the techniques and fundamentals employed by entrepreneurs to differentiate or diversify their businesses, reach business goals, and grow their businesses (Benson, 2017). For example, in Kenya, there is an increase in production facilities ideal for outsourcing, retail spaces for Kenyan-made fashion products, and a wider variety of materials. Entrepreneurs choose the type of business ownership based on its unique advantages and disadvantages.

Human capital/resources can make or break an enterprise. For excellent customer experience, the entrepreneur must first deliver great employee experience: reward competitively or fairly, train people in both skills and behaviours, and give them a stake in the business's success (Bindra, 2021b). The working conditions of manufacturing staff continue to focus on sustainable fashion research (Lantry, 2015). A system solves repeated business issues strategically and effortlessly (Benson, 2017).

It is vital that a start-up enterprise understands its competitors and takes measures to counter the competition using price and non-price strategies. Growth trajectory refers to critical aspects that will likely increase a business's development and take it to the next level. SMEs should grow organically with ethical, environmental, and socio-economic considerations (Benson, 2017). For instance, Angelsmile has diversified to ready-to-wear (RTW) to supplement its signature bespoke pieces (Robi, 2019). Outsourcing is both advantageous and disadvantageous to a firm. The former includes eliminating the need for a new business to install a sizeable in-house production infrastructure and freeing a designer entrepreneur to focus on the core business of designing (Agburu et al., 2017). Leather Masters, Frederick Bittiner Wear, and Panah allow fashion designers to produce RTW fashion products. Angelsmile fashion house wants to see Kenya's fashion sector become a glocalised and billion-dollar industry (Robi, 2019). Glocalisation is an amalgamation of the words globalisation and localisation, used to describe a service or product that is developed and distributed for the global market but is also designed and customised to accommodate the local users' culture.

Entrepreneurs are prone to err in decisions which negatively impact the enterprises. The impact is felt more by SMEs. Errors can occur in enterprise location, growth trajectory, financial and human capital management, and customer relations. A prudent entrepreneur not only addresses the errors but also learns from them.

Sustainability Principles and Design Business

Globally, entrepreneurs are implementing sustainability principles and ethos into their design processes. Specifically, the triple bottom line is an individual company's or entire value chain's ability to simultaneously satisfy profitability, environmental quality, and social justice (Lantry, 2015). Design is a determining factor in the success of products and services. Like other sustainable businesses, design businesses should engage in sustainable/cleaner production, considering the goods and services, workers, raw materials, capital, consumption, and communities (Sustainable, n.d.). For example, Manish Arora, an Indian fashion designer, collaborates with traditional artisans to create unique styles for a sustainable future for both actors.

Afrika's philosophy of Ubuntu, caring for others, is variously embodied by corporations. Giving back/conscience/corporate social responsibility (CSR) rise has created new types of collaboration and business benefits: financial and non-monetary. Several black female design entrepreneurs have extended their businesses to non-profit organisations (NPO) (Benson, 2017).

Social entrepreneurship (SE) is also gaining ground to give back to the community. A social enterprise is any business venture created for a social purpose and to generate social value while operating with a private-sector business's financial discipline, innovation, and determination. SE is fuelled by non-profits' quest for sustainability, mainly occasioned by declining support from traditional, philanthropic, and government sources and increasing competition for available funds (Padhiar & Garg, n.d.).

Business ethics are the moral principles/guidelines on how a business conducts itself and its transactions. Ethical issues are even more critical for start-ups and SMEs since their reputations are not as well established as big corporations. Typical business ethical issues include sexual harassment, nepotism, diversity and discrimination, social media, health and safety, environmental responsibility, accounting practices, and data privacy (Shabat, 2020). Ethical design entrepreneurs should commit to ethical practice.

Research Methodology

This chapter adopted a qualitative and case study research design. Three case studies of design businesses in Botswana and Kenya were purposively selected because they have more than five years in operation. The participants were in fashion and interior and graphic design. Purposive sampling was adopted to identify and select information-rich cases associated with the phenomenon of interest. The researchers contacted the SME entrepreneurs by email and telephone, requesting their participation in the study and permission/consent to use their images. Data collection adopted an interview schedule that was analysed using thematic analysis. Thematic analysis is "a method for identifying, analysing, and reporting patterns/themes within data" (Braun & Clarke, 2006, p. 79). The findings were presented in narratives of case studies and images.

Findings

The findings report on three companies sampled in the case study from Botswana and Kenya. Two enterprises were sampled from Botswana in interior design and graphic communication. In Kenya, a fashion design enterprise was sampled. The enterprises have been in business for over five years, and they managed to withstand the challenges that affected SMEs in the first two years. The enterprises were deemed good case studies that design students and other entrepreneurs can learn from for their own organisations and operations.

Interior Design Enterprise in Botswana

The enterprise is headed by a female entrepreneur aged 44 years, single and living in Gaborone, Botswana. She holds a Bachelor of Arts degree in Interior Design from American InterContinental University, USA. Additionally, she completed John Maxwell's leadership course.

Antecedents of Entrepreneurship

Entrepreneurship: The personality traits that spurred the entrepreneur to venture into entrepreneurship were passion, perseverance, and consistency. The hand-sketching skills she acquired in her undergraduate studies gave her the confidence to venture into entrepreneurship. This skill is pivotal to client communication. Her unique attributes are being focused, self-driven, patient, resilient, and taking challenges as a learning process. These attributes have sustained her through turbulent entrepreneurial times, thus ensuring the enterprise 18 years of operation.

Characteristics of SMEs and contribution to development agenda: The entity's characteristics include resilience and the ability to pivot and adapt to challenges encountered to remain relevant. Understanding one's market is critical. SMEs spur the development agenda considering their number within any economy and their ability to withstand economic turbulence. Further, they have the potential to upscale and create more employment.

The entrepreneur expressed the benefits of entrepreneurship, such as writing a script and being visionary. Further, an entrepreneur needs like-minded people to help realise the entity's vision. Human capital management is very challenging. Thus, the entrepreneur has learnt to be patient and a good listener. Financing has also been problematic, which would rather be a growth enabler. The enterprise has built a convincing profile that earned the trust of commercial banks.

Business profile: The enterprise is driven by a mission statement, vision, and core values aligned to the business's day-to-day running. The mentioned elements consistently remind them why they are in business and to reflect on their strides. The enterprise's mission "aims to provide professional services to the client's satisfaction by providing excellent service, honouring deadlines, and personalised design according to the client's requirement". The enterprise's vision is "delivering bespoke and innovative design solutions to the smallest details." The core values include a passion for creativity expressed in everything they do: customer-driven, seamless service delivery, bespoke design, and staff well-being. The firm offers interior design and project management services to the corporate, residential, retail, and hospitality industries. Manufacturing is done through its subsidiary and its homeware.

A total of ten employees on the design team are all qualified professionals holding Bachelor of Arts degrees in Interior Architecture from Botswana and Malaysia. The staff's key performance areas include developing conceptual designs, producing detailed drawings for production or construction, packaging design concepts for presentations, compiling project cost analyses, and project supervision. The employees are competent in using relevant software packages such as AutoCAD, ArchiCAD, Autodesk Revit, Adobe Photoshop, and Microsoft Office Suite. 3D rendering and good presentation skills employing various presentation media are critical. Soft skills needed are being a good team leader and being customer service oriented.

Business Strategies Adopted by SMEs

Dynamics for the business location, physical plant, and layout: The enterprise premises is in an up-and-coming commercial and residential area near supporting service providers. Having diversified to retailing, the location was selected to help brand the services and retail products. The physical layout ensures a joyous working space using colour, relaxation, breakaway areas, ergonomic furniture, and accessories.

Business life cycle phases and their attributes: The company has undergone three phases from its humble beginnings in a lounge. The growth and collaboration phases offered great opportunities to showcase its products and services through collaboration. Additionally, it had exciting moments, enabling the enterprise to be positioned in the corporate sector and hire interior designers. The third phase established a homeware section in 2020 during the COVID-19 pandemic as part of the expansion plan and complementing the interior design section. The homeware's core business is interior design homeware accessories, wallpaper, rugs, and, most importantly, manufacturing. The entity was non-committal on the firm's annual revenue.

The customer is king: The enterprise has been performing well in corporate services for the past ten years and intends to diversify into hospitality and retailing. Strategies to attract and retain clientele include bringing the clients on board into the project role, engaging the customer in every decision, and discussing risk and mitigation plans. Trust and professionalism are cemented, and clients become the firm's mouthpiece.

Know your competitor: The enterprise's competitors are young professionals who are technology savvy, which is an advantageous communication mode in the sector. The entity has competitors, and to remain competitive the enterprise offers reasonable space planning solutions, interrogates client requirements to ensure they cover all aspects, and ensures projects are always on budget and schedule. When clients need to articulate their requirements clearly, they probe further and advise. Nonetheless, the strategy sometimes works against them in tendering, as the competitors only offer minimal services, resulting in variations and extensions of the time during the project implementation.

The enterprise outsources visualisation and detailing services; thus, it can access advanced services, skills, and knowledge from experts that it cannot have permanently in-house. The greatest challenge in outsourcing is turnaround times. Nonetheless, the enterprise signs agreements with third parties on delivery turnaround times before the commencement of work, which manages expectations from both parties. The enterprise brand has a signature within its designs to celebrate diversity as a team. It fuses Afrikan elements that express to who they are as a people through patterns or accessories.

Business ownership model: The enterprise is owned by a father–daughter partnership. The father is a businessman and mentor. The partnership's advantage is the father mentoring the daughter; hence, it has not experienced enormous challenges.

Regarding growth trajectory, the enterprise was the first interior design company established in Botswana. The firm has grown exponentially, attributed to consistency and having the right team to work and collaborate with. Referral clients also spurred growth, indicating that the referee has done the tremendous groundwork. The turnover has also grown over the years, allowing the firm to sustain the business and diversify its portfolio. During the COVID-19 pandemic, the enterprise adopted cloud computing for remote access and continuity to serve its clients. The firm plans to put greater emphasis on visualisation, which needs to be improved. To sustain its growth, plans are underway to diversify its business to take control of all the critical project deliverable processes.

Human capital/resource management: The employees are recruited and selected from recommendations from institutions, referrals, and applications submitted. Strategies to retain employees include training, motivation/encouragement, and salary and benefits increments. The employer picks outstanding employees who are committed and loyal and supports them in the execution of their work. Notably, there is hardly any staff resignation. Thus, it does not suffer from the negative impact of high staff turnover.

Creativity is nurtured through brainstorming sessions: a platform for sharing ideas before a project commences. Designers are allowed to work individually on smaller projects. To motivate employees, the firm maintains an events calendar to celebrate each other's birthdays and calendar events. Employees need motivation because low staff morale affects employees and negatively impacts the business.

Unprecedented times: The COVID-19 pandemic made the enterprise adopt new technology: remote access from home, something it had been shying away from for a long time. Hence, staff continued working and delivering to the clients. For instance, they held weekly virtual meetings to set weekly targets. The firm was also forced to investigate residential projects for sustainability. Residential interiors are where the business started, and it was great to be back. The negative impact of the pandemic was low revenue with no ongoing projects.

Financial management: Initially, the enterprise was self-funded with the revenue it generated. Over the years, it has gotten funding from commercial banks. The firm has faced several challenges: as a start-up, commercial banks needed to trust such entities more easily. Female-led businesses also face economic discrimination. However, they have participated in interventions through the United Nations Development Programme to facilitate financing for women-led businesses and market access.

Going global: The enterprise hopes to expand outside Botswana through collaboration because the sector is saturated in that country due to a small market. The global market is attractive due to access to a more significant population.

To err is human: The entity has experienced failure to implement processes/manuals that determine employees' and employers' obligations in conducting business. The absence of such processes and guidelines opened myriad gaps in operating the business where employees manipulated the system. The lessons learnt from errors forced the business to formulate and implement processes and guidelines to govern their conduct and operations, such as employee manual and accounting processes. Mistakes are deemed as learning curves, but they should not be repeated.

Success factors in the business: refer to gratitude, satisfaction, and monetary reward. The firm's success is attributed to leadership and having a good team.

Sustainability Principles and Design Business

The enterprise defines sustainability as a contribution towards reducing carbon footprints. Sustainability principles adopted include procuring materials within Botswana where possible, and 60% of the materials chosen are made from recycled material or harvested responsibly. The entity views corporate social responsibility (CSR) as adding value to the less privileged communities. The firm adds value to children's learning and living environments. Engagement in CSR is not for benefits or recognition but to give back and show gratitude to the community they serve. The entity reported being a great believer in social entrepreneurship and should be encouraged. Professionals must share their skills and knowledge to improve their communities and the country.

Africa's philosophy of Ubuntu reminds us to always look back to whom we are as individuals and Afrikans, raised to be respectful to all. Ubuntu is embraced in the enterprise through encouraging respect and humanness towards one another and collaborative projects with local artisans.

Business ethics refers to conducting business with standard ethics and morals and without trampling on people's rights. The enterprise has encountered business ethical issues. From humble beginnings, the entrepreneur faced intimidation from male colleagues in meetings and sexual favours to advance the company's work. She stood steadfast to her principles and morals. Additionally, she experienced accounting malpractices that demanded management be more involved and vigilant.

Fashion Design Enterprise in Kenya

The owner/manager is a Kenyan woman aged 36 years, married, and residing in Nairobi. She holds a master's degree in Marketing Management from Middlesex University and was pursuing a PhD at the University of Nairobi, Kenya, at the time of writing. Additionally, she completed a professional course on business management.

Antecedents of Entrepreneurship

Entrepreneurship: The personality traits that spurred her to venture into entrepreneurship were tenacity, determination, and commitment. She identified a lack of variety in the fashion industry and a need in the bridal segment. The professional background prepared the owner/manager for design entrepreneurship – specifically, the ability to execute the technicalities required to understand and cut complex clothing. Her master's degree allows her to personally market the firm, which saves costs because outsourcing is expensive. The firm's ten years of operation in the unstructured situations of entrepreneurship are spurred by the owner/manager's resilience, creativity, openmindedness, and curiosity.

Characteristics of SMEs and contribution to development agenda: SMEs are flexible due to the ability to adapt to market changes, innovate, and allow for multitasking. Undoubtedly, SMEs contribute to job creation. Entrepreneurship is rewarding, pushes one to learn new challenges every day, leading to growth, and teaches responsibility and problem-solving. Nonetheless, entrepreneurship is a "lonely" and exhausting journey and requires multitasking.

Business profile: The enterprise lacks a mission and vision statement and core values. The total number of employees is eight, including the owner/manager. There are two degree holders, one diploma holder, three certificate holders, and a self-taught individual. Their job titles were operations manager, production assistant, assistant designer, and tailors.

Business Strategies Adopted by SMEs

Dynamics for the business location and physical plant and layout: The following factors were considered in selecting the firm's location: convenience for owner/manager and clients; sufficiently upmarket, affordable, and appreciated by the target clients; and parking availability. The physical layout offers comfort and luxury (within her capability) to make it feel like a bridal salon.

Business life cycle phases and their attributes: The firm has undergone two phases: start-up and growth. The former phase is experimental and exciting and requires lots of hard work to build a reputation. The growth phase faces increased competition, some fatigue, and the need for creativity to stay ahead. The owner/manager was non-committal on the firm's annual revenue.

The customer is king: Various strategies are deployed to attract and retain particularly middle- and upper-class clients, namely, consistent service, tirelessly responding to customers' needs, and maintaining good customer service as well as significant relationships with stakeholders (suppliers and clients, among others). Custom bridal gowns and accessories are the firm's mainstay, sold through custom orders and independent physical space (Figures 14.1–14.5).

FIGURES 14.1   2019 collections

(Photograph by Joan Mosomi)

FIGURES 14.2   2018 collections

(Photograph by Joan Mosomi)

Know your competitor: A competitor is anyone who sells a contemporary bridal gown, whether custom-made or ready-to-wear (RTW). The firm faces competition from other entities. To remain competitive, it researches new trends, sustains an active social media presence, and endeavours to maintain high standards. The firm outsources embroidery and some beadwork, thereby saving time and costs, focusing on core functions and reducing staffing needs. Outsourcing is challenging due to a need for more quality and output control. The brand reflects Kenya because it caters to Kenyan women's bodies and personal needs.

FIGURES 14.3 2017 collections

(Photograph by Joan Mosomi)

Business ownership model: The firm is a limited liability company with the advantage of having less risk but high taxation. The participant did not divulge the firm's growth trajectory.

Human capital/resource management: The employees are recruited and selected from industry recommendations and institutions where the

FIGURES 14.4   2016 collections

(Photograph by Joan Mosomi)

owner/manager teaches. In the past, she engaged a human resource company to assist with recruitment. Strategies to retain employees include fair
wages and treatment of workers, offering incentives, occasional rewards,
and providing training. Employee turnover destabilises business. The firm
nurtures employees' creativity by tasking them with taking on challenging

FIGURES 14.5 2015 collections

(Photograph by Joan Mosomi)

projects, collectively finding solutions, and endeavouring to take employees' suggestions on board. The employees are motivated by supporting them within and without (where possible) the work environment and paying them fairly/competitively. Motivating employees to ensure they give/do their best. Low morale among staff is occasioned by companies that make profit a priority above everything else. Poor work indicates low morale among employees.

Unprecedented times: The COVID-19 pandemic negatively affected the business, leading to loss of income and staff's low morale. The owner/manager saw no positive impact of COVID-19 on the business.

Financial management: The primary sources of financing/capital for the enterprise at the start-up phase were a loan from parents and savings. In the growth phase, finance is from organic growth and debt financing from HEVA – Africa's first dedicated finance, business support, and knowledge facility that funds the creative industries. Financial challenges emanate from the business's seasonality and difficulty in scaling up. The firm is still working on the solutions.

Going global: The entity did not have any plans of going global.

To err is human: Two mistakes in the entrepreneurial journey are over-reliance on team members and being too trusting, leaving the business vulnerable. Another mistake is not having proper systems in place. The firm is working on improving those constantly. The lesson learnt from the errors is the importance of installing systems right from the start of the business.

Success factors in business: Success refers to the firm's ability to run smoothly in the owner/manager's absence and to cater to the owner/manager's and team's needs. The enterprise's success story is attributed to being organised and consistent and having a unique product.

Sustainability Principles and Design Business

Though the enterprise did not define sustainability, the firm has adopted sustainability principles: limit or aim at zero waste – fair labour practices, work-life balance, good working conditions for the team, and equity. Sustainability benefits are low turnover, goodwill, and good customer service because of happy employees. Corporate social responsibility (CSR) is giving back, whereby the firm supports the education of a needy child in the team. Goodwill accrues from adopting CSR.

The enterprise needed to be more conversant with social entrepreneurship. The owner/manager neither defined nor vocalised the incorporation of Afrika's philosophy of Ubuntu in the enterprise. Business ethics refers to maintaining appropriate practices within one's business. The enterprise did not report any unethical business practices in the firm.

Graphic Communication Enterprise in Botswana

The enterprise was founded by three like-minded individuals who brought together their specialties and competencies to traverse the business world. The lead designer is married and holds a Bachelor of Design in industrial design and a business administration qualification.

Antecedents of Entrepreneurship

Entrepreneurship: Creative zeal spurred the participant into entrepreneurship. On the other hand, an observation of Botswana's graphic design scene led to the firm's establishment. The entrepreneur's professional qualification enabled accurate interpretation of customer needs and the generation of multiple concepts. Passion and resilience have enabled the firm to withstand and handle the unstructured situations of entrepreneurship. Entrepreneurship is beneficial because an individual executes her/his dream and passion and finally see it flourish. However, it is fraught with too many risks that may easily take someone off the course.

Characteristics of SMEs and contribution to development agenda: SMEs characteristically tend to lack resources. Nevertheless, they spur employment creation and sharing skills.

Business profile: Despite the entrepreneur not vocalising the firm's mission, vision statements, and core values, he is aware that they help communicate the company's intentions to employees and clients. The enterprise has operated for nine years with 15 staff members: nine permanent, four temporary, and two interns. The staff hold professional qualifications in design, ICT, accounting, and marketing. The firm offers targeted marketing through advertising, film, branding, and marketing services.

Business Strategies Adopted by SMEs

Dynamics for the business location and physical plant and layout: The location was selected to ensure easy access to the blue-chip client. The physical layout has an individual department housed in its own office within the building.

Business life cycle phases and their attributes: Currently, the firm is in the growth phase, which is characterised by uncertainties, namely seeking new and retaining old clientele, as well as trying out new ventures. The entrepreneur did not divulge the firm's annual revenue.

The customer is king: The enterprise's target clients are blue-chip companies and the government. The clients are attracted and retained by superior quality service.

Know your competitor: It is essential to know the competitors in any firm providing services within one's scope of business. Advertising and marketing agencies are the entity's competitors. The firm uses targeted marketing and superior-quality services to beat its competitors. Large-scale printing is outsourced, which enables the enterprise to punch above its height. Nonetheless, quality control is a challenge when outsourcing services. In terms of branding, the enterprise creates products with a feel of crafts that represent Botswana's culture.

Business ownership model: The enterprise is a partnership, thus endowed with multiple skill sets and quick decision-making. The main disadvantage is the partners' financial insolvency. Regarding growth trajectory, the entity started from zero to several clientele, from the city outskirts to the city centre, and from one to the current 15 employees. It is considering branching out into other spheres and beefing up machinery to maintain its growth.

Human capital/resource management: The enterprise advertises posts and relies on the probation period to assess a new employee. Employees are offered workplace incentives to stay. The firm adheres to its policy on employee separation. Employer turnover contributes to increased costs of employee training and can lead to variations in adapting to the organisation's culture. Therefore, the entity nurtures the employees' creativity through periodical employee retreats involving team building and "Friday idea explosions", where they sit around and bounce ideas off each other. The staff are motivated through its intrapreneurship programme that allows some employees to own a certain percentage of the business revenue they attracted to the company. Motivating employees to help build a strong brand that delivers on its promises is critical. In turn, the effort automatically leads to a high-performing culture that increases revenue for the business and a high turnover for employees' remuneration and shareholders' dividend. Employees' low morale is attributed to the non-existence of a unique organisational culture that all employees should be inducted into from the beginning. Indicators of low staff morale include low-performance turnout and high employee turnover.

Unprecedented times: The positive impact of the COVID-19 pandemic on business was the adoption and use of technology, namely virtual workspace. However, most businesses cut down their advertising budgets. Thus, low revenues were experienced.

Financial management: Financial conundrums relate to cash flow crunch and access to a larger pool of capital for expansion. The solution is to apply for external funding from angel firms and institutions that rally behind Afrikan entrepreneurs, namely the Tony Elemenu Foundation.

Going global: The enterprise defines a global SME with a high growth potential that can be incorporated into a global player. The firm has yet to venture outside Botswana.

To err is human: Lack of financial planning has taught the entity always to have accessible cash flow and maintain tight controls over all financial activities.

Success factors in business: Success means delivering on its brand promise and getting value from it. The entity's success is pegged on its expansion into an array of other industries, namely property.

Sustainability Principles and Design Business

The enterprise defines sustainability as being conscious and considerate of the social, economic, and environmental impacts/outcomes caused by the company's activities. The firm has adopted gender equality and clean technology (UV printing). The latter has resulted in increased revenue. Regarding CSR, the firm adheres to open communication and offers practical industry experience to students from different institutions of higher learning. The benefits accruing from adopting CSR include helping to raise awareness on sustainability adherence as well as raising industry-ready graduates. Social entrepreneurship is specifically focused on social, cultural, and environmental outcomes. However, the enterprise does not operate as a social enterprise. Africa's philosophy of Ubuntu addresses humanity, a core value amongst all people of Afrikan descent. Thus, the enterprise has embraced Ubuntu by being conscious and disapproving of all materials/services and products that involve child slavery and human trafficking. Business ethics means moral conduct in business. The firm has not experienced any unethical practices.

Discussion

Different design businesses have a place in Afrika, be it communication, fashion, or interior design. The enterprises withstood the unstructured entrepreneurship environment to operate the design enterprises for 9–18 years. The finding contradicts Greeff (2019) that most South Afrikan SMEs fail within the first two years. The enterprises' longevity could be attributed to mainly professional design and business-related qualifications and the pursuit of further studies to enrich the management of the business. Business-related training includes marketing, leadership, business administration, and family business management. Though the sample was small, the result diverges from Rapitsenyane et al. (2019), who established a prevalence of non-fashion-related qualifications among the fashion houses' owner-managers/founders: Master of Urban and Regional Planning, accounting, and biology, that could suggest a lack of background knowledge and skills in fashion design and technical dynamics in the industry. The entrepreneurs established the firms in their youth; when one is most active, the absorptive capacity is high, coupled with high creativity, and this also enabled them to engage in up-skilling.

Antecedents of Entrepreneurship

The enterprises' survival is also credited to entrepreneurs' personality attributes and implementation of superior business strategies. Passion is the main driver for venturing into the creative industry, in addition to creative zeal, tenacity, determination, commitment, perseverance, and consistency, significantly because entrepreneurship is predominantly hands-on, requiring long hours. Thus, without passion, one will quit within no time. The entrepreneurs also conducted extensive market dynamics research to find a niche – notably, a lack of locally owned interior design firms and a need in the bridal segment.

The women entrepreneurs surmounted female-oriented conundrums, specific intimidation by male colleagues and prospective clients seeking sexual favours in exchange for business, and financial instability. This finding correlates with Mwendapole et al. (2013) that Botswana's female designers experience gender disparities, tender awards, and clients prefer doing business with male designers regardless of their capability. Cho et al. (2021) also allude to gender stereotypes and access to funds among women entrepreneurs in South Korea. These challenges are contrary to Botswana's idiom that says "*mosadi thari ya Sechaba*" (a woman ties the nation on her back). Nevertheless, women uphold the culture and heritage (Grant & Grant, 1995). On the contrary, Robb and Watson (2012) argue that there is no difference between genders in entrepreneurial performance.

The proprietors capitalise on the peculiar characteristics of SMEs especially being adaptive to changes in the market, allowing for multitasking, offering opportunities for innovativeness, and creating meaningful employment for between eight and ten workers. The finding concurs with Mehta (2018) that MSMEs spur development by creating employment, providing desirable sustainability and innovation (new products, services, or techniques), and promoting competitiveness. Further, amidst intense competition from multinational companies, SMEs have been resilient and fiercely fight for their turf (Drucker, 1985).

Business Strategies Adopted by SMEs

The entrepreneurs deem their entrepreneurial ventures successful, epitomised by diversification into other sectors, such as real estate/property and homeware established during the COVID-19 pandemic, unique products, being organised and consistent, high-level leadership, and having a good team.

Regarding business strategies, the entrepreneurs opted for upmarket business space in capital cities that would appeal to their targeted clientèle – upper and middle class, blue-chip corporates, and government, as well as affordability and proximity to service providers. The result echoes Barnard et al. (2011) that the business location directly affected the business's survival rate. Most successful SMEs were in the city centre, usually awash with resources/business inputs (Greeff, 2019). The longevity of the design firms is also credited to the entrepreneurs' cognisance that human resources/capital is the cog of a successful firm. The entrepreneurs endeavour tirelessly to value, motivate, nurture creativity, and retain the staff, which has seen a low turnover of employees. Low employee turnover accords firm stability and mitigates the punitive cost of 60% employee turnover (Bindra, 2021a).

Despite the global disruption occasioned by ICT and the COVID-19 pandemic, the participants still employ the traditional selling methods of targeted marketing, custom orders, and independent physical space, probably due to the low uptake of digital services in Afrika. Non-price strategies are predominantly deployed to outwit competitors.

To overcome the lack of machinery and/or technical know-how in some processes, the entrepreneurs engage in open innovation/outsourcing, enabling them to punch above their height, accruing benefits and challenges. All organisation forms are outsourced regardless of company size (Isaksson & Lantz, 2015). Nonetheless, outsourcing reduces quality control, meeting deadlines, and ethical and environmental compliance (Lantry, 2015).

The businesses are growing organically from start-up to the current growth and maturity phases, despite encountering hurdles. The exponential growth is manifested in increased clientèle, relocating premises, recruiting more employees, and outsourcing. The result echoes Benson (2017) that the growth trajectory includes hiring a team, location, manufacturing, and wholesaling.

Different phases/life cycles of a business require unique sources of finance to stay afloat. The start-up phase is significantly financed from personal savings and loans from individuals who believe in the entrepreneur's dream. During the growth phase, the funding sources diversify to include organic growth, credit from commercial banks and debt financing institutions. However, a significant challenge for SMEs is a limited financial muscle, especially for female-led businesses. A notable finding is that all the entrepreneurs were mum about their financial standings, which could be attributed to their inability to manage finances strategically.

The entrepreneurs humbly admitted having erred in business operations and learnt from their mistakes. Among the mistakes is not implementing transparent systems/manuals/processes in the infancy phase, which could have slowed the firm's growth and development. Benson (2017) asserts that creating systems allows the fashion design entrepreneur time to focus on core business aspects. The entrepreneurs' systems related to production, transactions, conduct, and obligations of all actors within and outside the firm and availed for smooth running.

Sustainability Principles and Design Businesses

The proprietors have an in-depth understanding of the global sustainability discourse and practically implement the principles in their processes and operations: environmental, economic, and cultural sustainability. The latter is evidenced by adopting Afrika's philosophy of Ubuntu. The entrepreneurs have taken up the call of Sustainable Measures (n.d.) that design businesses should engage in sustainable/cleaner production. Several benefits accrue from entrepreneurs' sustainability initiatives, significantly increased revenue from using current technologies such as UV printing; relatively low employee turnover; goodwill; and good customer service because of happy employees. The entrepreneurs' implementation of CSR paybacks includes raising awareness of sustainability adherence and supporting the design profession by raising industry-ready graduates and goodwill. According to Benson (2017), the business benefits of CSR, among other things, are attracting and retaining key and talented employees, more robust clients, community relations, and improving a firm's financial health. The lack of understanding about SE has inhibited all the proprietors from establishing a social enterprise. Benson (2017) asserts that several entrepreneurs have extended their fashion design business to include non-profits. SE is fuelled by NPOs' quest for sustainability (Padhiar & Garg, n.d.). The same would apply to business sustainability. The entrepreneurs' in-depth comprehension of the philosophy of Ubuntu promotes its infusion into the products and services offered and successful collaboration with other actors, namely artisans. Lantry (2015) espouses that successful collaborations can be genuinely inspiring and allow artisans to gain sustained employment as equal partners in the co-design and co-execution of any initiative.

Conclusion and Recommendations

This chapter dispels the notion that business acumen and design know-how are mutually exclusive. Notably, Afrika is replete with successful design businesses in diverse sectors. This study provides insights into Afrika's design entrepreneurship from the perspectives of three proprietors: two females and one male, whose businesses have existed for up to 18 years, defying the fifth anniversary limit for most SMEs. The success is attributed to a mix of dynamics: entrepreneurs' attributes, professional and business-related qualifications, and robust business strategies adopted. The enterprises are continually making an immense socio-economic contribution to their respective country's development, even in the face of the COVID-19 pandemic and gender-based conundrums. This study illustrates how women design entrepreneurs have navigated the male-dominated space, mainly interior design, to successfully operate their firms and are still going strong. The women entrepreneurs are more on stylistic, high-end luxury that reflects who they are and their design ethos, which is felt from their inspirational words. An intentional implementation of a sustainable product-service system by entrepreneurs shall offer several benefits related to dematerialisation, resource efficiency, system-oriented attributes, and sustainability by offering services such as maintenance, renting, upgrading, redesigning, swapping, or lending (Rapitsenyane et al. (2019).

The following recommendations were made based on the findings:

1 Design industry players must establish design hubs within higher education institutions offering design courses and hosting industry captains to support aspiring designers, where they will be incubated before progressing to entrepreneurial space. The effort shall create industry-ready and entrepreneurial-minded graduates.
2 Policy intervention and programmes are crucial in Afrika to address gender-based conundrums experienced by women and SMEs, among them design-oriented SMEs. The interventions could include tax incentives, access to finance and markets, and entrepreneurship mentorship.
3 Customised design entrepreneurship courses must be included in curricula in higher education institutions to impart the students with business acumen that shall be implemented upon graduation.

Although this study provided an overview of design entrepreneurship in Afrika, further research must be undertaken to uncover successful design enterprises specialising in *inter alia* graphic, soft/textile furnishings, textile, industrial, and leather design.

Reference List

Agburu, J. I., Anza, N. C., & Iyortsuun, A. S. (2017). Effect of outsourcing strategies on the performance of small and medium scale enterprises (SMEs). *Journal of Global Entrepreneurship Research*, 7(1), p. 26.

Barnard, S., Kritzinger, B., & Krüger, J. (2011). Location decision strategies for improving SMME business performance. *Acta Commercii*, 1(1), pp. 111–128.

Benson, S. L. K. (2017). *Black Fashion Designers' Matter: A Qualitative Study Exploring the Experiences of Black Female Fashion Design Entrepreneurs*. Doctorate thesis, Ames: Iowa State University Capstones, United States of America.

Bindra, S. (2021a, November 7). Not holding on to staff? You might want to think about hidden costs. *Sunday Nation*, p. 35.

Bindra, S. (2021b, November 14). To offer excellent customer experience and deliver great employee experience. *Sunday Nation*, p. 31.

Braun, V., & Clarke, V. (2006). Using thematic analysis in psychology. *Qualitative Research in Psychology*, 3(2), pp. 77–101.

Brush, C. G., Greene, P. G., & Welter, F. (2020). The Diana project: A legacy for research on gender in entrepreneurship. *International Journal of Gender, and Entrepreneurship*, 12(1), pp. 7–25.

Cho, Y., Park, J., Han, S. J., Sung, M., & Park, C. (2021). Women entrepreneurs in South Korea: Motivations, challenges and career success. *European Journal of Training and Development*, 45(2/3), pp. 97–119. https://doi.org/10.1108/EJTD-03-2020-0039.

Delavelle, F., & Rouanet, L. (2017). *Female Entrepreneurship Key Ingredient for Africa's Growth*. https://ideas4development.org/en/ffemale-entrepreneurship-key-ingredient-africa-growth/.

Drucker, P. F. (1985). *Innovation and Entrepreneurship: Practice and Principles*, 1st ed. Oxford: Elsevier.

Ghosh, P. K., Ghosh, S. K., & Chowdhury, S. (2018). Factors hindering women entrepreneurs' access to institutional finance an empirical study. *Journal of Small Business and Entrepreneurship*, 30(4), pp. 279–291.

Grant, S., & Grant, E. (1995). *Decorated Homes in Botswana*. Mochudi, Botswana: Phuthadikobo Museum.

Greeff, M. (2019). *Product Design Challenges of Small to Medium Enterprises in the Western Cape: A Design Thinking Approach*. Masters thesis, Cape Town: Cape Peninsula University of Technology.

Isaksson, A., & Lantz, B. (2015). Outsourcing strategies and their impact on financial performance in small manufacturing firms in Sweden. *International Journal of Business and Finance Research*. 9(4), pp. 11–20.

Lantry, J. (2015). *Rethinking Sustainability Through Collaborative Exchange Between Emerging Australian Designers and Indian Artisans in Fashion and Textiles*. Masters thesis, Sydney: University of Technology Sydney. https://www.academia.edu/27034859/Artisan_Culture_Rethinking_Sustainability_through_Collaborative_Exchange_between_Emerging_Australian_Designers_and_Indian_Artisans_in_Fashion_and_Textiles.

Mehta, S. (2018). *Design at the Doorstep: Design Approaches for MSMEs*. Ahmedabad: National Institute of Design. www.nid.edu.

Muriithi, S. M. (2017). African Small and Medium Enterprises (SMEs) contributions, challenges, and solutions. *European Journal of Research and Reflection in Management Sciences*, 5(10), pp. 36–48.

Mwendapole, C., Mapfaira, H., Setlhatlhanyo, K., & Kutlwano, P. (2013). The business of design: The emerging field of design entrepreneurship in Botswana. *GIDEC 2013 Conference*, 23–26 September, Gaborone, Botswana.

Nasar, A., Akram, M., Safdar, M. R., & Akbar, M. S. (2021). A qualitative assessment of entrepreneurship amidst the COVID-19 pandemic in Pakistan. *Asia Pacific Management Review*, 27(3), pp. 182–189.

Nzohabonimana, D. (2016). The creative economy in Africa. *Rwandair Inzozi Magazine*, December 2015–February 2016, pp. 70–72.

Padhiar, V., & Garg, R. (n.d.). *Convergence of Culture and Economic Empowerment – A Multiple Case Study of Women Artisans in the Handicraft Sector of Gujarat*. www.academia.edu.

Rapitsenyane, Y., Njeru, S., & Moalosi, R. (2019). Challenges preventing the fashion industry from implementing sustainable product service systems in Botswana and Kenya. In A. Gwilt, A. Payne, & E. A. Rüthschilling (Eds.), *Global Perspectives on Sustainable Fashion*, pp. 236–245. London: Bloomsbury Visual Arts.

Robb, A. M., & Watson, J. (2012). Gender differences in firm performance: Evidence from new ventures in the United States. *Journal of Business Venturing*, 27(5), pp. 544–558.

Robi, L. (2019). *Meet Angelsmile's Wambui Kibue, the Game Changer in the Kenyan Fashion Industry*. www.newssummedup.com.

Shabat, B. (2020). *8 Common Ethical Issues Facing Businesses in 2021*. www.become.com.

Smith, A., Young, A. A., & Raeside-Elliot, F. (2015). Teaching business concepts using visual narrative. In R. VandeZande, E. Bohemia, & I. Digranes (Eds.), *LearnxDesign. Proceedings of the 3rd International Conference for Design Education Researchers*, pp. 1552–1568, p. 4. Chicago. www.academia.edu.

Sustainable Measures. (n.d.). *Definitions of Sustainable Business and Production*. www.sustainablemeasures.com.

United Nations. (2015). *Sustainable Development*. www.sdgs.un.org/goals.

Yang, H., Park, J., & Seol, B. (2016). An exploratory study on the characteristics of knowledge and technology-based women entrepreneurship in Korea. *Innovation Studies*, 11(1), pp. 113–141.

15

DESIGNING FUTURES FOR AFRIKA AND BEYOND

Vikki Eriksson, Keineetse Christopher Motlhanka, and Thatayaone Mosepedi

Introduction

The ability of the design process to address complex challenges is well documented. 'Wicked problems', a phrase defined and described by Rittel and Webber (1973), aimed to capture the complexity of various social challenges that could not be solved with a single, linear way of thinking. Instead, an iterative way of engaging, bringing together different ways of thinking, would be needed. Buchanan (1992, p. 20) offered insight into how products should be viewed in this frame:

> As a humanistic art of technological culture, design strive towards a new attitude about the appearance of products. Appearance must carry a more profound, integrative argument about the nature of the artificial in human experience. This argument is a synthesis of three lines of reasoning: the ideas of designers and manufacturers about their products, the internal operational logic of products, and the desire and ability of human beings to use products in everyday life in ways that reflect personal and social values. Effective design depends on the ability of designers to integrate all three lines of reasoning.

The view on more integrative ways of thinking and addressing complex challenges kept evolving the industrial design profession. The authors believe this evolution is still ongoing. From an Afrikan perspective, designing from the ethos of Ubuntu is a design-thinking strategy that embraces indigenous knowledge systems and acknowledges the Afrikan voice. With the future of design education in mind, Tan (2009) noted that a designer's role had evolved

DOI: 10.4324/9781003270249-21

and functioned as a co-creator, researcher, provocateur, social entrepreneur, facilitator, capacity builder, and strategist. The range of emerging skills indicated that designers are no longer merely involved in making but also in understanding, reflecting, actioning, and commenting. These skills will underpin the future practice of industrial designers, with an inevitably changing role of toggling between the drawing board and prototyping facilities and the board rooms. This strategic role embodies the potential of design as a transdisciplinary practice to harness technology and place the human experience, quality of life, and the environment central to development. Speculative approaches can offer a critical lens through which the future can be imagined, and the impact of people's consumption and lifestyles can be recognised. These evolving approaches to thinking and practice within the context of design will enable designers to explore the immense challenges facing Afrika, and the world, today.

With the global influence of climate change and social inequalities, the impact of a design needs no longer be localised. Afrikan solutions to inclusive policy development, sustainable product creation, improved health and wellness systems, and global connectivity will be helpful across the globe. The Afrikan continent, and individual Afrikan countries, represent great diversity in culture, language, living circumstances, and environments. As such, designs from these regions often embody a sensitivity to contexts and empathy. Working within the parameters of limited resources or infrastructure has also imbued Afrikan design with an inherent resilience and responsiveness. These critical components of future designs require designers to be more mindful of resources, impact, consumption, and humanity. In addition to evolving philosophies and methodologies, a future-focused design must embrace new materials and various production methods, from high-tech to artisanal and from single units to mass production.

The Changing Role of Designers

As a noun, the word can describe a drawing, prototype, or concept representing an idea or solution to a specific challenge or question. As a verb, the word implies an iterative investigation, sense-making, creation, and evaluation process. The lack of a definitive definition for the word *design* has historically been one of the challenges when framing design as a professional practice. Creative practitioners use the same word to describe *what* and *how* they do it. The openness of design as a term is mirrored by the ever-evolving and changing nature of the word 'designer'.

It can be argued that an industrial designer in the initial part of the 20th century focussed more on the artefact. As the century progressed, industrial designers explored the seemingly endless possibilities that material innovation and technological advances allowed. Products from Europe and North

America were brought to Afrika by those who colonised the continent and indigenous travellers. The creative industries of Afrika were showcased in Britain and Europe at trade shows and exhibitions (such as the 1924 Wembley Empire Exhibition). However, these exhibitions often failed to recognise the level of sophistication of Afrikan products and their innovative nature. They did contribute to international awareness of Afrikan design philosophies, materials, aesthetics, and production methods (Woodham, 1998). Embedded in Afrikan product creation and products from many global indigenous populations was a focus on sustainable material use and production methods (Grant et al., 2018; Walker, 2013). These products were also not created with planned obsolescence, a tenet of the philosophy of Ubuntu.

Towards the end of the 20th century, the consequences of mass production and product consumption became clear. Even before this, the impact on the planet was undeniable, but the critical design voices of individuals like Fuller (1969) and Papanek (1972) brought the severity of the situation to the attention of a broader audience. These authors critiqued mainly Western design practices that had led to a world filled with unnecessary products and normalised a 'throw away' culture in which items were quickly replaced, not repaired, or updated. Society had to recognise the role of products in their lives and the consequences of their unsustainable consumption behaviour. Designers, who had contributed to the plethora of products on the market that fuelled material culture, began using design to explore the complex challenges society faces. Many of the challenges aim to establish a more sustainable way of living. Modern designers must play an active role in creating products and services that acknowledge the importance of sustainability and harness technological innovations to better the lives of all people, expanding their skills to include a range of functions that were considered outside the scope of design in previous centuries.

Future Skills and Competence: The Future Industrial Designer

Industrial designers are not the only practitioners experiencing a constantly evolving world of work. Design is closely linked to human activity; designers experience rapid changes more instantly. As future technologies emerge, designers need to find ways to activate them for improved products, services, and systems. As new environmental and social challenges present, designers must respond and find appropriate solutions. Designers have emerged as key players in shaping the future. Creative problem-solving, collaboration, and other skills that one may associate with design are often cited as critical future skills. Whiting (2020), commenting on the World Economic Forum's (2020) Future of Jobs Report, noted that required skills in the future workplace would include complex problem-solving, ideation, creativity, originality, self-management, and critical thinking. The fluency of ideas is another essential future skill, defined as the ability to come up with several different ideas when presented with a challenge (Bakhshi et al., 2017). It is through design that these skills, and many others, can be developed. Working iteratively and engaging in a designerly way of thinking fosters 21st-century skills and allows the practitioner to experience empathy for others (Noweski et al., 2012).

Designerly thinking represents one of the most critical assets of future industrial designers. It allows one to consider and investigate challenges systematically. It allows testing and experimentation and encourages input and collaboration with many stakeholders. This is critical, as machine learning enables a higher level of product personalisation (Abdu et al., 2020). An exciting opportunity within the customisation and personalisation of products is the ability to create products that a diverse group can use.

As industrial designers aim to collaborate more closely with individuals during the process to ensure a more 'people-centred' design outcome, they will function as mediators, researchers, capacity builders, and facilitators. Projects may be so complex that the skills of a range of professionals are needed during the design process, which requires designers to function as communicators and even project managers. Moving into the future, the knowledge and skills required of an industrial designer will remain fluid, adapting to leverage new technologies and expanding to ensure responsible choices are made. Mastering new technologies and techniques that support the design process will require industrial designers to be comfortable working at a micro (individual product or design feature) and macro (systems and contexts of engagement and use) level.

Systemic design offers practitioners an integrated approach to design for large, social, complex challenges. Working in this manner allows for proposed design interventions that acknowledge all individuals who are stakeholders and non-human influences that must be considered. Industrial designers are critical stakeholders in approaches like service design and systemic design. Many physical connections are required to make services accessible; think of a mobile phone or app interface as an access point to mobile services or a banking website and ATM as access points to financial services. Due to the complexity of services and systems, these designs must be a collaborative practice between relevant individuals and specialists. As the need for working in this manner increases, multi and transdisciplinary engagement has become essential.

The design of future Afrikan products and services must focus not only on the traditional criteria of ergonomics, aesthetics, accessibility, technical specification, and engineering and sincerely acknowledge those who will use them, they must also acknowledge the systems and interconnected nature of the modern world. What designers create today will impact tomorrow and shape the future. As technology enables more complex solutions, multi- and transdisciplinary groups of individuals offer more significant insights and a broad scope of understanding when seeking possible solutions.

Transdisciplinary Practice: Designers as Collaborators

Transdisciplinary design approaches effectively address large-scale and complex problems innovatively by transcending disciplinary boundaries from multiple angles to address the need for a more holistic approach to problem-solving in contemporary Afrika. When advocating for transdisciplinarity, all stakeholders need to ensure their participation in the problem-solving process goes beyond inter- and multidisciplinary methods, as these lack the necessary collaboration and synthesis accorded by transdisciplinarity. Traditional cross-disciplinary approaches are limited by their occurrence as interactions between different disciplines. At the same time, transdisciplinary activities require full disciplinary integration, which unifies all disciplines and their knowledge base, resulting in 'new thinking' and 'new autonomous bodies of knowledge' (Leblanc, 2007; Ertas, 2018). Designers must project themselves as transdisciplinary leaders competent at bringing diverse disciplines together.

Afrikan challenges present the perfect illustration of complex problems requiring designers to think beyond improving the look (aesthetic) and functionality of products, which often results in increased redundancy. The use of digital manufacturing technology, especially in rapid prototyping (RP), is fast garnering a reputation as a leader in demonstrating effective collaboration between design and other disciplines. One such example is the collaboration between the Central University of Technology (CUT) in South Afrika and a team of surgeons to design and produce 'fully sensate and functioning limbs in which specialised implants have replaced resected bone' instead of the more traditional prosthetic limbs which are produced using additive manufacturing (Truscott et al., 2007). The collaboration of designers with medical practitioners allowed surgeons to involve their patients more in developing prosthetics and other medical products by collecting patient-specific data through 3D scanning capabilities and using CAD models. This allowed the patients to give feedback on less technical issues, such as aesthetic preferences, to ensure designs meet specific patient needs. Doctors could also rely on additive-manufactured physical models to plan for complicated medical procedures, thereby shortening theatre operation times and reducing associated costs.

Humanising Technology: Focussing on Experience and Interaction

The relationship between people and technology has fascinated technologists, designers, researchers, sociologists, anthropologists, psychologists, and others interested in future and development studies. Two critical concepts for the future of design are acknowledging that a broad range of individuals must be considered when imagining future products and services and that technological advances cannot be seen as separate from the development of individuals and society.

Technology can connect individuals to critical resources, products, and services and plays a vital role in enabling different work practices. To facilitate and improve the way people interacted with the different forms of technology, the process of engagement and touchpoints had to be considered in the first half of the 20th-century studies aimed to optimise how people worked and tried to explore the conditions needed for success. These studies recognised that technology influenced and shaped the ability of individuals to be productive in the same way that individuals shaped technology, implying a reciprocal nature. Fields and processes such as human–computer interaction (HCI), user experience (UX) design, user interface (UI) design, and interaction design (IxD) have continued to emphasise the relationship between the person using the product or service and the technology which enables it.

For industrial designers, UX design methods and processes allow the exploration of individual experiences of products, systems, or services. IxD focuses on the design of interactive products, services, systems, and environments; these can be digital, analogue-physical, or a mix. Beyond considerations of optimisation and personalisation, these approaches offer designers the opportunity to be more inclusive in engagements. Technology's design possibilities must be crafted into responsive and empathetic product solutions. Afrika is a vast continent, and different Afrikan communities have unique requirements for products and services. This can be linked to physical design requirements, such as environmental considerations (for example, terrain considerations if the goal is to design for mobility), material considerations, semiotics, and leveraging local production and knowledge systems. Afrikan industrial designers must fulfil the role of custodians and advocates of sustainability, relevance, accessibility, and diversity during the design process. Design approaches that celebrate the collaboration with stakeholders and focus on engagement and interaction can yield the research that makes this possible.

Technology in Afrika: Overcoming the Digital Divide and Embracing the 4IR

The effects of the Fourth Industrial Revolution (4IR) in developed countries have resulted in significant developments in communication – social media platforms like Facebook, WhatsApp, and LinkedIn – and manufacturing – evident through the rapid product development of products such as cell phones, gaming consoles, etc. These effects trickle down into developing countries where the outcomes of the 4IR are less apparent in environments still struggling to catch up with the second and third Industrial Revolutions. Despite reports of immense global growth in internet usage from Afrikan countries, only 24% of people on the continent accessed the internet, a stark difference from the 80% of people in Europe against the 51% global average of internet users (Markowitz, 2019).

Lessons from previously missed opportunities in developing economies make it clear that Afrikan countries cannot afford to fall behind once again in the current technological growth, as adopting these new technologies will prove critical for rapid and sustained productivity and economic growth. Developing economies, Afrika included, stand to benefit from the significant growth being experienced by the general purpose technologies driving the 4IR or 'Industry 4.0', which include big data, the Internet of Things (IoT), artificial intelligence (AI), 3D printing, cloud computing, robotics, advanced materials, genetics, nanotechnology, and biotechnology through increased productivity in agricultural and manufacturing activities, development of new products and new markets, the emergence of new economic activities, as well as exponential growth in employment in both the formal and informal sectors (Markowitz, 2019; Ayentimi, 2020).

Most Afrikan economies are still driven by agriculture and natural resource extraction, activities not primarily considered essential for the implementation of emerging technologies associated with the 4IR and while this is rightly so, it remains crucial for Afrikan countries to strike a balance between rapid adoption of new technologies and development of new industries and adapting them to growing and developing capabilities in these economic activities. In Botswana and other Afrikan countries, the economic importance of livestock to rural livelihoods and foreign exchange earnings from exporting to the European Union has resulted in the implementation of disease control measures, such as the erection of cordon fences (Kock et al., 2006). In Zimbabwe, Mudziwepasi and Scott (2015) explored the use of wireless sensor networks (WSN) to develop a remote cattle monitoring system to be used as an assistive technology known as zero-effort technologies (ZETs) to link authorities to herds of cattle and enable them to monitor their health and movement. In East Afrika, radio frequency identification (RFID) technology has been used in combination with the internet, short messaging (SMS), and GPRS technologies to aid government officials found along the cross-border regions between Kenya, Uganda, Sudan, Ethiopia, and Somalia in the identification and recovery of stolen cattle. This problem has consumed the area for a long time resulting in conflict and instability among the many inhabitants of the region. For Afrika to attain successful levels of the 4IR implementation, all stakeholders must work together to overcome the continent's challenges.

Design for the Internet of Things

The growth of the 4IR is primarily driven by the significant progress in the development of general-purpose technologies, including the Internet of Things (IoT), which is being spurred on by research from companies and institutes constantly and rapidly developing new technologies, standards, platforms, applications, and devices. As a technology, the IoT is considered the most disruptive technology currently after the World Wide Web and universal mobile accessibility. It gives people the capability to use devices to reach out into the world from any location, making use of advances in RFID, short-range wireless communications, real-time localisation, and sensor networks (Feki et al., 2013).

Industrial design is experiencing a transformation from being product-centric, i.e., physical products not connected, to those that are more experience-centric, i.e., connected products that provide the user with an overall experience. The shift from product-centric to experience-centric design has resulted in the growth of specialist design disciplines like user UX and IxD. The upsurge in the development of products for IoT has resulted in increased demand for UX and interaction designers in developed economies, with an estimated 13% increase between 2010 and 2020 (Murray, 2013; Interaction Design Foundation, 2020). UX and interaction designers focus their design processes on complex ecosystems capable of collecting data during use and relaying this information to other devices and then processing the information using cloud computing, where designers can access the information and use it accordingly (Chen, 2012) to improve both the products and the customer experience.

Designers developing products and services for IoT in Afrika need to be cognizant of the areas that have already used IoT to address societal issues of high importance, such as vehicle tracking, monitoring of air quality in Kenya, and the health sector where an HIV diagnosis communications system allows for test results from far away labs to reach clinics quicker and to save lives in the process (Ndubuaku & Okereafor, 2015). Central to successful design in Afrika will be the ability to produce and test prototypes of designs with real end users and leverage new production opportunities.

Changing Production: Rapid Prototyping, 3D Printing, and the Future of Manufacturing

Weller et al. (2015) observe that additive manufacturing, rapid prototyping, or 3D printing, is currently being promoted as the spark of the 4IR. Go and Hart (2016) further describe rapid prototyping as expanding the design space for complex objects, enabling the exploration of new material properties, and changing how designers configure local and global supply chains. Rapid prototyping has been commonly known to be a process used solely for prototyping and testing. This is because most of its techniques were incapable of large-scale production in a cost-effective way. However, the latest technological advances, materials, and other crucial build aspects like production speed, cost, and tolerance have made it a sought-after option. The latest prototyping used as a design technique has been critical in realising unique and functional models.

Rapid prototyping allows the manufacturing industry to produce geometrically complex, low- to medium-volume production components in a range of materials with substantially compressed lead time (Wong et al., 2014). As digital transformation influences reduction in lead times, the desire for sophisticated products and the use of other types of prototyping approaches will decline. Rapid prototyping and 3D printing are revolutionary smart manufacturing techniques and key terms within the global digital parameters. They remain the future mainstream manufacturing technology for industrial designers in Afrika for Industry 4.0. The impacts of Industry 4.0 mean that future manufacturing techniques and the world will revolve around high digital technology. Industrial designers with computer-aided design (CAD), computer-aided manufacturing (CAM), and 3D printing skills set will identify with Industry 4.0 and be able to serve the related industries of the future.

These smart manufacturing technologies can contribute to solving Afrikan challenges that include the adverse impact of climate change, increasing water scarcity, biodiversity and ecosystem loss, desertification, low resilience to natural disasters, energy crisis, food crisis, limited benefits from globalisation, health security, the global financial crisis, trafficking and piracy, low penetration of ICT services, urbanisation, need to develop better disaster response mechanisms, genetically modified crops, food security, and technology transfer among others (United Nations Economic Commission for Africa, 2015).

This challenge has made it difficult for Afrika to keep pace with digital transformation and migration like the rest of the world. Smart manufacturing is a reality, representing a new stage in manufacturing and control of the industrial value chain. Its adoption has caused much disruption in developed countries. Considering the points mentioned earlier, Afrika is still challenged to catch up with the rest of the world in embracing the digital future, which is required for readiness to adopt smart manufacturing and embrace Industry 4.0.

Considering Materials for Tomorrow's Design Practice

Afrika is endowed with a wealth of natural resources. However, raw materials are exported to developed countries often without any value addition. Design can contribute to developing the value chains of the exported raw materials regionally to benefit local communities. At a local level, raw materials can be processed to be compatible with smart manufacturing techniques like rapid manufacturing, often used solely for prototyping and testing. Nevertheless, with the latest technological advances, materials production cost, and tolerance, rapid manufacturing has been critical in realising unique and functional prototypes. It also provided the industry with the capabilities to produce geometrically complex, low- to medium-volume production components in various materials with sustainable compressed lead times (Wong et al., 2014).

Prototyping materials such as polyurethane foam and cardboard, usually used for low-fidelity prototypes, often help evaluate form and usability issues such as the reach and positioning of controls.

In addition to reductions in lead times, more recent development in digital manufacturing technologies has been in printing on various materials such as clay (Revelo & Colorado, 2018) and cementitious materials (Carneau et al., 2020). Using high technology to produce prototypes quickly for customers will become an important and convenient business for various industries, especially with the opportunities that 3D printing presents to distributed manufacturing (DM) (Srai et al., 2016). This is especially true with the rise of open-source designs for mass customisation and manufacturing (Schelly & Pearce, 2019). These advantages of smart manufacturing will make them the most appropriate materials of choice in Industry 4.0.

Smart manufacturing, however, does not use traditional everyday manufacturing materials. A set of novel materials ranges from liquids to powders and sheets. As smart manufacturing evolves, it will dictate the pace for the discovery and use of future manufacturing materials. Smart manufacturing techniques will follow their path and be open to all materials, including organic-based materials and biomaterials (Kusiak, 2018). Four-dimensional printing (4D) is one of the latest offerings in the manufacturing sector that is fast gathering pace. It enables the fabrication of smart materials, such as methacrylate-based thermos-responsive polymers, with an adaptation of shapes and properties (Han et al., 2021). Conducted studies show that 4D offers a significant reduction in total part cost as opposed to 3D. Using stimuli-responsive shape memory material (SMMS) has enabled programmable self-adaptation of printed structures concerning their shape and properties when exposed to pre-determined external stimuli (Han et al., 2021). Future design and technology will impact the physical and material worlds and the augmented and virtual worlds.

Future Realities: Design for the Virtual, Augmented, and Mixed

A new technology landscape is emerging in the form of human-assisted artificial intelligence (AI), augmented reality (AR), virtual reality (VR), and mixed reality (MR), referred to as reality-virtuality technologies (Flavián et al., 2019) which requires designers to be aware of and be able to adjust their processes to evolve with the digital environment. These new technologies are predicted to play critical roles in several industries, such as design, mechanical, military, retail, tourism, education, medical and healthcare, entertainment, and research (Mei et al., 2019).

Future realities as proponents for the IoT provide product designers with increased volumes of data informing design decisions in ways that enable designers to develop personalised user experiences using dynamic user interfaces which analyse user interactions with products. AR use has been predominantly used on smartphones and tablets, but designers are now exploring expanding this to other products such as smart glasses; product designers and UX designers will have to explore new design processes to develop what has traditionally been a physical product into a more intelligent and smart device capable of communicating with the user when in use and constantly collecting user data to assist them in making every decision, which could include which routes to take when travelling, and weather forecasts for the day. Developed economies are exploring using these technologies for assisted and enhanced living applications. Judith Okonkwo, founder of Imisi 3D, a company focused on developing AR/VR solutions in Nigeria, explains the potential AR, VR, and MR technologies must address issues around education and healthcare, two of the most pressing issues in the continent. VR as a learning tool can enable students in low-resourced schools to carry out experiments virtually and access virtual site visits to places they would typically not be able to travel to (Chiefe, 2019). Digital technology has transcended beyond existing in computers, smartphones, and tablets in the form of apps and websites; there is a growing desire from customers for devices which provide interactive, physical-virtual connections to engage with users on a deeper level to be able to communicate with them consistently for different types of assistance. Industrial designers need to be aware of this transition from product-centric to experience-centric and develop products that will meet the future's needs.

Conclusion

It is fitting if designers want to work in a way that is environmentally sustainable and socially sensitive, the design must:

> be radical and revolutionary in the truest sense. It must dedicate itself to nature's principle of least effort, in other words, maximum diversity with minimum inventory. . . . This means consuming less, using things longer, and being frugal about recycling materials . . . [what] designers can bring to the world must now be combined with a sense of social responsibility. In many areas, designers must learn how to redesign. In this way, we may have survived through design.
>
> *(Papanek, 1985, pp. 346–347)*

Afrikan industrial design is truly the future of global industrial design. As different technologies become mainstream and offer ways to improve the quality of life of individuals and communities, designers will play an integral role in moulding them to the needs of people and the planet. The globe is more connected than ever, with modern communities forming outside traditional geographic proximity. Indeed, the next version of life is expected to be a metaverse, an environment that blends physical and digital realities (Lee, 2021). The emergence of this environment will blend various technologies and extended realities and offer new design opportunities to leverage the possibilities of the metaverse (Lik-Hang et al., 2021).

The early phases of hyper-connected environments can be found in smart homes and cities worldwide. Connected through the Internet of Things, products and appliances are shifting from isolated, function-driven objects to an extended mesh of interconnected nodes. It is essential that Afrikan designers and technologists contribute to the shaping of the metaverse, as Zoghlami (2021) notes: in Afrika, people no longer want to be only consumers; they want to be actors, to influence, to embrace technology, and to think about the future as well as that of the entirety of humanity. Active participation of stakeholders in design decision-making is a tenet aligned with the philosophy of Ubuntu. If new technologies can adopt the philosophy of Ubuntu, the acceptance rate will be enhanced.

For industrial designers, the opportunities that technology presents will manifest both within external environments, such as our homes and places of work, and alter our physical bodies. Bionic limbs and other body augmentation products can address limb loss or form part of human restorative development. Wearable technology pushes socio-technical interaction boundaries, external devices, or permanent features within people's bodies.

Designers, however, have the responsibility to look beyond the allure of the new and critically engage with the contextual relevance, accessibility, and cost of emerging technologies. Cost implies the financial cost to the individual purchasing the final product and the social and environmental cost of the product's production, distribution, use, and disposal. The design of products and services to support emerging and established green technologies such as electric mobility, renewable energy production and storage, artificial photosynthesis, biomimetic design, and carbon capture is critical. Mastering the process of activating these technologies means little if they cannot be transformed into usable devices. The future products, systems, and services in Afrika must place the well-being of individuals central in the design process and resist producing wasteful or redundant items, thus not sacrificing the planet's future or people.

Reference List

Abdu, S., Romouzy, A., & El-Yazed, A. (2020). The future of Industrial design in view of machine learning. *International Design Journal*, 24(2), pp. 115–127. https://doi.org/10.21608/idj.2020.81510.

Ayentimi, D. T. (2020). The 4IR and the challenges for developing economies. In *Developing the Workforce in an Emerging Economy*, pp. 18–30. Oxfordshire: Routledge. https://doi.org/10.4324/9780429273353.

Bakhshi, H., Downing, J. M., Osborne, M. A., & Schneider, P. (2017). *The Future of Skills: Employment in 2030*. London: Pearson Publishing.

Buchanan, R. (1992). Wicked problems in design thinking. *Design Issues*, 8(2), pp. 5–21. https://doi.org/10.2307/1511637.

Carneau, P., Mesnil, R., Roussel, N., & Baverel, O. (2020). Additive manufacturing of cantilever-from masonry to concrete 3D printing. *Automation in Construction*, 116, p. 103184. https://doi.org/10.1016/j.autcon.2020.103184.

Chen, Y. K. (2012). Challenges and opportunities of the internet of things. *Proceedings of the Asia and South Pacific Design Automation Conference*, ASP-DAC, pp. 383–388, October 2012. https://www.researchgate.net/publication/254019549_Challenges_and_opportunities_of_internet_of_things.

Chiefe, U. (2019). Five insights on the current state of virtual reality in Africa. *Techpoint Africa*. https://techpoint.africa/2019/08/16/virtual-reality-africa/.

Ertas, A. (2018). *Transdisciplinary Engineering Design Process*. New Jersey: John Wiley & Sons.

Feki, M. A., Kawsar, F., Boussard, M., & Trappeniers, L. (2013). The internet of things: The next technological revolution. *Computer IEEE*, 46(2), pp. 24–25. https://doi.org/10.1109/MC.2013.63.

Flavián, C., Ibáñez-Sánchez, S., & Orús, C. (2019). The impact of virtual, augmented, and mixed reality technologies on the customer experience. *Journal of Business Research*, 100, pp. 547–560. https://doi.org/10.1016/j.jbusres.2018.10.050.

Fuller, R. B. (1969). *Operating Manual for Spaceship Earth*. Carbondale: Southern Illinois University Press.

Go, J., & Hart, A. J. (2016). A framework for teaching the fundamentals of additive manufacturing and enabling rapid innovation. *Additive Manufacturing*, 10, pp. 76–87. https://doi.org/10.1016/j.addma.2016.03.001.

Grant, E., Greenop, K., Refiti, A. L., & Glenn, D. J. (Eds.). (2018). *The Handbook of Contemporary Indigenous Architecture*. New York: Springer.

Han, M., Yang, Y., & Li, L. (2021). Techno-economic modelling of 4D printing with thermo-responsive materials towards desired shape memory performance. *IISE Transactions*, pp. 1–13. https://doi.org/10.1080/24725854.2021.1989093.

Interaction Design Foundation. (2020). The basics of user experience design BY interaction design foundation. *The Basics of User Experience Design*, pp. 21–27. www.interaction-design.org/ebook.

Kock, M. D., Mullins, G. R., & Perkins, J. S. (2006). *Wildlife Health, Ecosystems, and Rural Livelihoods in Botswana*. New York: Oxford University Press.

Kusiak, A. (2018). Smart manufacturing. *International Journal of Production Research*, 56(1–2), pp. 508–517. https://doi.org/10.1080/00207543.2017.1351644.

Leblanc, T. (2007). Transdisciplinary design approach. In P. Kotze et al. (Eds.), *Creativity and HCI: From Experience to Design in Education*, pp. 106–122. Aveiro, Portugal: Springer.

Lee, J. Y. (2021). A study on metaverse hype for sustainable growth. *International Journal of Advanced Smart Convergence*, 10(3), pp. 72–80. https://doi.org/10.7236/IJASC.2021.10.3.72.

Lik-Hang, L. E. E., Braud, T., Zhou, P., Wang, L., Xu, D., Lin, Z., Kumar, A., & Hui, P. (2021). *All One Needs to Know about Metaverse: A Complete Survey on Technological Singularity, Virtual Ecosystem, and Research Agenda*. https://arxiv.org/abs/2110.05352v3.

Markowitz, C. (2019). Harnessing the 4IR in SADC: Roles for policymakers occasional paper 303. *South African Institute of International Affairs*, October, pp. 1–47. www.weforum.org/agenda/ [Accessed 10 June 2022].

Mei, Y., Nie, Q., Wang, F., Lin, Y., & Jiang, H. (2019). Application of augmented reality technology in industrial design, IOP conference series. *Materials Science and Engineering*, 573(1), p. 012062. https://doi.org/10.1088/1757-899X/573/1/012062.

Mudziwepasi, S. K., & Scott, M. S. (2015). Assessment of a wireless sensor network based monitoring tool for zero effort technologies: A cattle-health and movement monitoring test case. *IEEE International Conference on Adaptive Science and Technology*, ICAST, pp. 1–6, January 2015, IEEE, Ota, Nigeria.

Murray, T. (2013). *How the Internet of Things Is Changing Industrial Design – Bresslergroup, Bressler Group*. www.bresslergroup.com/blog/how-the-internet-of-things-is-changing-industrial-design/.

Ndubuaku, M., & Okereafor, D. (2015). Internet of things for Africa: Challenges and opportunities. *Proceedings of International Conference on Cyberspace Governance: The Imperative for National & Economic Security*, pp. 22–31, 4–7 November 2015. https://www.researchgate.net/publication/287997186_Internet_of_Things_for_Africa_Challenges_and_Opportunities.

Noweski, C., Scheer, A., Büttner, N., von Thienen, J., Erdmann, J., & Meinel, C. (2012). Towards a paradigm shift in education practice: Developing twenty-first-century skills with design thinking. In *Design Thinking Research*, pp. 71–94. Berlin, Heidelberg: Springer. https://doi.org/10.1007/978-3-642-31991-4_5.

Papanek, V. (1972). *Design for the Real World: Human Ecology and Social Change*. New York: Pantheon Books.

Papanek, V. (1985). *Design for the Real World: Human Ecology and Social Change*, 2nd ed. Chicago: Academy Chicago.

Revelo, C. F., & Colorado, H. A. (2018). 3D printing of kaolinite clay ceramics using the Direct Ink Writing (DIW) technique. *Ceramics International*, 44(5), pp. 5673–5682. https://doi.org/10.1016/j.ceramint.2017.12.219.

Rittel, H. W. J., & Webber, M. M. (1973). Dilemmas in a general theory of planning. *Policy Sciences*, 4(2) pp. 155–169.

Schelly, C., & Pearce, J. M. (2019). Policies to overcome barriers for renewable energy distributed generation: A case study of utility structure and regulatory regimes in Michigan. *Energies*, 12(4), p. 674. https://doi.org/10.3390/en12040674.

Srai, J. S., Kumar, M., Graham, G., Phillips, W., Tooze, J., Ford, S., & Tiwari, A. (2016). Distributed manufacturing: Scope, challenges, and opportunities. *International Journal of Production Research*, 54(23), pp. 6917–6935. https://doi.org/10.1080/00207543.2016.1192302.

Tan, L. (2009). Perspectives on the changing role of the designer: Now and to the future [Paper presentation]. *ICSID Design World Design Congress Education Conference, Design Education 2050*, 23–25 November 2009, Singapore: Temasek Polytechnic.

Truscott, M., De Beer, D., Vicatos, G., Hosking, K., Barnard, L., Booysen, G., & Campbell, R. I. (2007). Using RP to promote collaborative design of customised medical implants. *Rapid Prototyping Journal*, 13(2), pp. 107–114. https://doi.org/10.1108/13552540710736795.

United Nations. Economic Commission for Africa. (2015–03). *Economic Report on Africa 2015: Industrializing Through Trade*. Addis Ababa, UN: Economic Commission for Africa. https://hdl.handle.net/10855/22767.

Walker, S. (2013). *The Handbook of Design for Sustainability*. London: A&C Black.

Weller, C., Kleer, R., & Piller, F. T. (2015). Economic implications of 3D printing: Market structure models in light of additive manufacturing revisited. *International Journal of Production Economics*, 164, pp. 43–56. https://doi.org/10.1016/j.ijpe.2015.02.020.

Whiting, K. (2020). These are the top 10 job skills of tomorrow – and how long it takes to learn them. *World Economic Forum*. www.weforum.org/agenda/2020/10/top-10-work-skills-of-tomorrow-how-long-it-takes-to-learn-them/.

Wong, D. S., Zaw, H. M., & Tao, Z. J. (2014). Additive manufacturing teaching factory: Driving applied learning to industry solutions. *Virtual and Physical Prototyping*, 9(4), pp. 205–212. https://doi.org/10.1080/17452759.2014.950487.

Woodham, J. (1998). Images of Africa and design at the British Empire exhibitions between the wars. *Journal of Design History*, 2(1), pp. 15–33. www.jstor.org/stable/1315746.

World Economic Forum. (2020). *The Future of Jobs Report*. https://www3.weforum.org/docs/WEF_Future_of_Jobs_2020.pdf?_gl=1*140sxrk*_up*MQ.&gclid=Cj0KCQjwhY-aBhCUARIsALNIC05wqlhEEdxLT-eZkCorOYfe-M0UBl4fHBYX7SE590cnZbxRW9T-T5EkaAuNEEALw_wcB.

Zoghlami, M. (2021). Cart'Afrik: Will Africa embrace the metaverse? *African News Agency*. www.africanewsagency.fr/cartafrik-will-africa-embrace-the-metaverse/?lang=en.

INDEX